McDougal Littell
Algebra 1
Concepts and Skills

Larson   Boswell   Kanold   Stiff

CHAPTER 3   # Resource Book

The Resource Book contains the wide variety of black-line masters available for Chapter 3. The blacklines are organized by lesson. Included are support materials for the teacher as well as practice, activities, applications, and assessment resources.

McDougal Littell
A HOUGHTON MIFFLIN COMPANY
Evanston, Illinois   •   Boston   •   Dallas

Contributing Authors

The authors wish to thank the following individuals for their contributions to the Chapter 3 Resource Book.

Rita Browning
Linda E. Byrom
José Castro
Rebecca S. Glus
Christine A. Hoover
Carolyn Huzinec
Karen Ostaffe
Jessica Pflueger
Barbara L. Power
James G. Rutkowski
Michelle Strager

ISBN: 0-618-07853-3

456789-DWI-04 03 02

Contents

3 *Solving Linear Equations*

Contents

Contents

Descriptions of Resources

This Chapter Resource Book is organized by lessons within the chapter in order to make your planning easier. The following materials are provided:

Tips for New Teachers These teaching notes provide both new and experienced teachers with useful teaching tips for each lesson, including tips about common errors and inclusion.

Parent Guide for Student Success This guide helps parents contribute to student success by providing an overview of the chapter along with questions and activities for parents and students to work on together.

Prerequisite Skills Review Worked-out examples are provided to review the prerequisite skills highlighted on the Study Guide page at the beginning of the chapter. Additional practice is included with each worked-out example.

Strategies for Reading Mathematics The first page teaches reading strategies to be applied to the current chapter and to later chapters. The second page is a visual glossary of key vocabulary.

Lesson Plans and Lesson Plans for Block Scheduling This planning template helps teachers select the materials they will use to teach each lesson from among the variety of materials available for the lesson. The block-scheduling version provides additional information about pacing.

Warm-Up Exercises and Daily Homework Quiz The warm-ups cover prerequisite skills that help prepare students for a given lesson. The quiz assesses students on the content of the previous lesson. (Transparencies also available)

Activity Support Masters These blackline masters make it easier for students to record their work on selected activities in the Student Edition.

Alternative Lesson Openers An engaging alternative for starting each lesson is provided from among these four types: *Application, Activity, Graphing Calculator,* or *Visual Approach.* (Color transparencies also available)

Graphing Calculator Activities with Keystrokes Keystrokes for four models of calculators are provided for each Technology Activity in the Student Edition, along with alternative Graphing Calculator Activities to begin selected lessons.

Practice A and B These exercises offer additional practice for the material in each lesson, including application problems. There are two levels of practice for each lesson: A (transitional) and B (average).

Contents

Reteaching with Practice These two pages provide additional instruction, worked-out examples, and practice exercises covering the key concepts and vocabulary in each lesson.

Quick Catch-Up for Absent Students This handy form makes it easy for teachers to let students who have been absent know what to do for homework and which activities or examples were covered in class.

Learning Activities These enrichment activities apply the math taught in the lesson in an interesting way that lends itself to group work.

Interdisciplinary Applications/Real-Life Applications Students apply the mathematics covered in each lesson to solve an interesting interdisciplinary or real-life problem.

Challenge: Skills and Applications Teachers can use these exercises to enrich or extend each lesson.

Quizzes The quizzes can be used to assess student progress on two or three lessons.

Chapter Review Games and Activities This worksheet offers fun practice at the end of the chapter and provides an alternative way to review the chapter content in preparation for the Chapter Test.

Chapter Tests A and B These are tests that cover the most important skills taught in the chapter. There are two levels of test: A (transitional) and B (average).

SAT/ACT Chapter Test This test also covers the most important skills taught in the chapter, but questions are in multiple-choice and quantitative-comparison format. (See *Alternative Assessment* for multi-step problems.)

Alternative Assessment with Rubrics and Math Journal A journal exercise has students write about the mathematics in the chapter. A multi-step problem has students apply a variety of skills from the chapter and explain their reasoning. Solutions and a 4-point rubric are included.

Project with Rubric The project allows students to delve more deeply into a problem that applies the mathematics of the chapter. Teacher's notes and a 4-point rubric are included.

Cumulative Review These practice pages help students maintain skills from the current chapter and preceding chapters.

Cumulative Test This is a quarterly test that can be used to assess students' progress on the key skills and concepts in Chapters 1–3.

LESSON 3.1

TEACHING TIP Some students will be able to solve easy one-step equations mentally or by using their calculator. Nevertheless, be consistent in writing the solution steps required for each equation and ask students to do the same when they are completing their work. Although students might think that this is a waste of their time, they must get into the habit of showing all solution steps now. Otherwise they will not be able to solve multi-step equations later.

COMMON ERROR Students might have difficulties with equations where the variable is subtracted from a number, such as $4 - x = 5$. Some of them will add 4 to both sides of the equation because they see the minus sign between the 4 and the x. Point out that the important sign or operation is the one *in front of* the number they want to "move." In this case, they must get rid of 4, which can also be written as $+4$. They do this by subtracting 4 from both sides of the equation because subtraction undoes addition. Students must also realize that after the 4 is gone, they have to solve an equation with a negative variable, namely $-x = 1$.

LESSON 3.2

COMMON ERROR Some students might be confused as to what to do with equations where a negative number is multiplied by a variable, such as Example 1 on page 138. They might try to add 4 to both sides of the equation to get rid of the -4. To help these students, read the exercise to students as "negative 4 times x is 1." Then ask them what operation is in this equation and how they can undo it. Make sure that they divide by -4, instead of by 4, by stressing the distinction between the operation and the sign of the number involved in it.

LESSON 3.3

TEACHING TIP Students get confused with multi-step equations because they do not know what to undo first. Make a connection to order of operations, which they already know. When solving an equation you are "undoing" things, so

you must also reverse, or undo, the order of operations. Therefore, addition and subtraction take priority over multiplication and division. The relationship between solving equations and the order of operations is like the one between opening and wrapping a present: you must undo what you did in the reverse order you followed to do it.

COMMON ERROR Some students will make mistakes combining like terms, such as changing $4 - 4x + 3x = -2x - 2$ to $4 - 7x = -2x - 2$. Review combining like terms.

LESSON 3.4

TEACHING TIP Ask students to solve an equation twice, first by collecting the variable terms on the right side, and then on the left side. Note that the answer will be the same either way. Discuss why they should collect the variable terms on the side with the greatest variable coefficient.

LESSON 3.5

INCLUSION Students with limited English proficiency or learning disabilities may have difficulties that are non-mathematical. You can help these students by reading the problems out loud and explaining any vocabulary words unknown to them. Then ask students to circle any "key words" or data that might be used to solve the problem. (Remind students to update their list of key words.) Next, ask them to tell you in their own words what the problem says. Only when you are satisfied that students comprehend the problem can you start writing a verbal model, labels, and an algebraic model.

LESSON 3.6

INCLUSION Review place value by writing a large decimal number and asking students which digit is in the tenths, thousandths, millionths, and so on.

LESSON 3.7

TEACHING TIP When solving formulas, first ask students to think what order they would follow to evaluate the formula if they knew the values to plug in. Then, remind them that solving involves undoing the operations in the reverse order. In Example 1 on page 171 where $C = \frac{5}{9}(F - 32)$, if you knew the value of F you would subtract 32 first and then multiply by the fraction. Therefore, to solve for F you must first eliminate the fraction and then add 32.

TEACHING TIP Another method to help students solve formulas involves replacing all variables with numbers except for the one we want to solve for. Write the resulting equation and the original formula side by side and solve each of them by taking the same steps. By using this method, students can easily transfer their knowledge of solving equations to solving formulas. For instance:

$$40 = \frac{5}{9}(F - 32) \qquad C = \frac{5}{9}(F - 32)$$
$$\frac{9}{5} \cdot 40 = \frac{9}{5} \cdot \frac{5}{9}(F - 32) \qquad \frac{9}{5} \cdot C = \frac{9}{5} \cdot \frac{5}{9}(F - 32)$$

COMMON ERROR When working with functions, some students might say that if $y = \dfrac{12 - 2x}{3}$, then $y = 4 - 2x$. Review how to use the distributive property, rewriting the function as $y = \frac{1}{3} \cdot (12 - 2x)$ and drawing arrows to show how to distribute the fraction.

LESSON 3.8

TEACHING TIP Some students might be confused with problems such as Example 6 on page 179. The problem can also be set up as two ratios. The first ratio is the unit analysis: $\dfrac{9.242 \text{ pesos}}{1 \text{ dollar}}$.

The second ratio is set up so the variable, x, is the number of pesos you're trying to find, over the total number of dollars. The students may need to be reminded that the units must be consistent, numerator to numerator, denominator to denominator. That is, pesos to pesos, dollar to dollar. After setting the two ratios equal to each other, the students will cross multiply to find x.

LESSON 3.9

INCLUSION Start with very easy percent problems with students who might have difficulties reading wordy problems. You might want to make sure to cover one of each of the three types of percent problems, namely, finding the percent, finding the initial total amount, and finding the amount after the percent is taken. Identify key words that students can use as a reference to write their equations.

TEACHING TIP Percent problems can also be solved as two ratios that are equal. To set up the ratios, just remember "is over of" and that the percent must be written as out of 100. In this manner, a problem such as "What is 25% of 72?" can be set up as $\dfrac{25}{100} = \dfrac{x}{72}$. Similarly, 16 is what % of 78 will be set up as $\dfrac{16}{78} = \dfrac{x}{100}$. Finally, 15 is 20% of what number will be set up as $\dfrac{15}{x} = \dfrac{20}{100}$.

Outside Resources

BOOKS/PERIODICALS
Nord, Gail D. and John Nord. "An Example of Algebra in Lake Roosevelt." *Mathematics Teacher* (February 1995); pp. 116–120.

ACTIVITIES/MANIPULATIVES
Hand-on Equations. Balance and pieces to represent and solve linear equations. Vernon Hills, IL; ETA.

SOFTWARE
Grade Builder: Algebra 1. Cambridge, MA; The Learning Company.

VIDEOS
Algebra in Simplest Terms. Linear equations. Burlington, VT; Annenburg/CPB Collection, 1991.

Parent Guide for Student Success

For use with Chapter 3: Solving Linear Equations

Chapter Overview One way that you can help your student succeed in Chapter 3 is by discussing the lesson goals in the chart below. When a lesson is completed, ask your student to interpret the lesson goals for you and to explain how the mathematics of the lesson relates to one of the key applications listed in the chart.

Lesson Title	Lesson Goals	Key Applications
3.1: Solving Equations Using Addition and Subtraction	Solve linear equations using addition and subtraction.	• Temperature Changes • City Parks
3.2: Solving Equations Using Multiplication and Division	Solve linear equations using multiplication and division.	• Restoring Movies • Thunderstorms
3.3: Solving Multi-Step Equations	Use two or more steps to solve a linear equation.	• Firefighting
3.4: Solving Equations with Variables on Both Sides	Solve equations that have variables on both sides.	• Steamboats • Gazelles and Cheetahs
3.5: More on Linear Equations	Solve more complicated equations that have variables on both sides.	• Astronomy • Rock Climbing • Computer Time
3.6: Solving Decimal Equations	Find exact and approximate solutions of equations that contain decimals.	• Retail Purchases • Cocoa Consumption • Fundraising
3.7: Formulas	Solve a formula for one of its variables.	• Mars Pathfinder • Scuba Diving
3.8: Ratios and Rates	Use ratios and rates to solve real-life problems.	• Bald Eagles • Gasoline Mileage • Exchange Rate
3.9: Percents	Solve percent problems.	• Electoral Votes • Choosing a College • Retail Purchases

Study Tip

Make Formula Cards is the study tip featured in Chapter 3 (see page 130). Encourage your student to make a card for each formula and to include a sample problem on each card. You can use the formula cards to help your student review the formulas and prepare for quizzes and tests.

NAME _____ DATE _____

Parent Guide for Student Success

For use with Chapter 3: Solving Linear Equations

Key Ideas Your student can demonstrate understanding of key concepts by working through the following exercises with you.

Lesson	Exercise
3.1	Write and solve an equation to answer the question. On a shopping trip, Clara and Sean Robinson's mother spent the same amount of money as Clara and Sean combined. If Clara spent $37 and their mother spent $64, how much did Sean spend?
3.2	Write and solve an equation to answer the question. One fifth of the students attending Ayla Middle School play sports. If 136 students play sports, how many students attend the school?
3.3	Solve the equation. $25 = 3(5x - 1) - x$
3.4	Solve the equation if possible. $\frac{2}{3}(12 - 9y) = 2(7 - 4y)$
3.5	Write and solve an equation to answer the question. Steven Spitz can buy either a CD player for $189 and CDs for $9.75 each or a cassette player for $216 and cassettes for $5.25 each. For what number of CDs or cassettes do the two options cost the same if Steven wants to buy the same number of CDs or cassettes?
3.6	Solve the equation. Round the result to the nearest hundredth. $14.8 + 3.94x = 2.47x - 12.8$
3.7	Rewrite the equation so x is a function of y. Then use the result to find x when $y = -3$. $7y - 3(x + 4) = 18$
3.8	The speed limit on the interstate expressway in many states is 65 miles per hour. How many kilometers per hour is that? (1 mi = 1.609 km)
3.9	When Courtney ran for student council president, she won 55% of the votes. If 198 people voted for her, how many people voted altogether?

Home Involvement Activity

Directions: Find the cost for two long distance phone services, including any monthly fees and the charge per minute for calls within the U.S. Let x be the number of minutes of long distance service used in a month. Write an equation and solve it, if possible, to find for how many minutes the two services cost the same. Which service is cheaper if you call fewer than this number of minutes? Which service is cheaper for your family?

Answers
3.1: $x + 37 = 64$; $27 **3.2:** $\frac{1}{5}x = 136$; 680 students **3.3:** 2 **3.4:** 3 **3.5:** $189 + 9.75x = 216 + 5.25x$; 6 CDs or cassettes **3.6:** -18.78 **3.7:** $x = \frac{7}{3}y - 10$; -17 **3.8:** about 104.6 km/hr **3.9:** 360

NAME _____ DATE _____

Prerequisite Skills Review

For use before Chapter 3

EXAMPLE 1 *Finding Opposites and Reciprocals*

Find the opposite and the reciprocal of the number.

a. 2 **b.** -3 **c.** $\dfrac{1}{4}$

SOLUTION

	Opposite	Reciprocal
a. 2	-2	$\dfrac{1}{2}$
b. -3	3	$-\dfrac{1}{3}$
c. $\dfrac{1}{4}$	$-\dfrac{1}{4}$	4

Exercises for Example 1

Find the opposite and the reciprocal of the number.

1. 5 **2.** $-\dfrac{2}{3}$ **3.** $\dfrac{1}{5}$

4. -7 **5.** -1 **6.** 10

EXAMPLE 2 *Checking Possible Solutions*

Check whether the numbers -6 and 2 are solutions of the equation $7 + 3y = -11$.

SOLUTION

To check the possible solutions, substitute them into the equation. If both sides of the equation have the same value, then the number is a solution.

y	$7 + 3y = -11$	Result	Conclusion
-6	$7 + 3(-6) = -11$?	$-11 = -11$	-6 is a solution.
2	$7 + 3(2) = -11$?	$13 = -11$	2 is not a solution.

Therefore, the number -6 is a solution of $7 + 3y = -11$.

The number 2 is not a solution of $7 + 3y = -11$.

Exercises for Example 2

Check whether the given number is a solution of the equation.

7. $4a - 6 = 18; -3$ **8.** $-42 = -5x + 3; 9$

9. $9 - 8y = 17; -1$ **10.** $6x - x = 11.01; 2.2$

NAME _____ DATE _____

Prerequisite Skills Review

For use before Chapter 3

EXAMPLE 3 *Using the Distributive Property*

Use the distributive property to simplify the expression.

a. $-4(2y + 17)$ **b.** $(6 - x)(-2)$

SOLUTION

a. $-4(2y + 17) = (-4)(2y) + (-4)(17)$ Using the distributive property, multiply each term by -4.

$\qquad\qquad\qquad\quad = -8y - 68$ Simplify.

b. $(6 - x)(-2) = (6)(-2) + (-x)(-2)$ Using the distributive property, multiply each term by -2.

$\qquad\qquad\qquad\quad = -12 + 2x$ Simplify.

Exercises for Example 3

Use the distributive property to simplify the expression.

11. $8(12 - y)$ **12.** $(2x - 45)(-3)$

13. $-5.8x(x - 2.5)$ **14.** $\frac{3}{7}(7z - 3y)$

EXAMPLE 4 *Simplifying Like Terms*

Simplify the expression.

$5(2x - 3) - 4x$

SOLUTION

$5(2x - 3) - 4x = 5(2x) + 5(-3) - 4x$ Use distributive property.

$\qquad\qquad\qquad\quad = 10x + 15 - 4x$ Multiply.

$\qquad\qquad\qquad\quad = 6x + 15$ Combine like terms.

Exercises for Example 4

Simplify the expression.

15. $2x + 3(x + 1)$ **16.** $4(m + 7) - m$ **17.** $2(3x + 2x - y)$

18. $4y + 6y + 10y$ **19.** $-5(x - 10)$ **20.** $-4(3y - n) - 2y$

Strategies for Reading Mathematics

For use with Chapter 3

Strategy: Reading Examples

Algebra books are filled with examples to help you understand the many ways
in which a concept can be applied. During your first reading of an example,
make sure you understand what it is trying to show you. Check that you follow
each step. Later, you can scan the examples to find ones that will help you
solve particular problems.

EXAMPLE 6 *Using a Known Formula*

Each example, like
this one from your
text, has a specific
purpose. As you
solve a problem,
read the appropriate
example.

Checking Vital Signs A body temperature of 95°F or lower may indicate the
medical condition called hypothermia. What temperature in the Celsius scale
may indicate hypothermia?

The Fahrenheit and Celsius scales are related by the equation $F = \frac{9}{5}C + 32$.

SOLUTION

Use the formula to convert 95°F to Celsius.

Examples may not show
each step. Make sure you
understand what was
done in any missing step.

Check that you
understand why
each number is
substituted.

$F = \frac{9}{5}C + 32$ **Write known formula.**

$95 = \frac{9}{5}C + 32$ **Substitute 95° for F.**

$63 = \frac{9}{5}C$ **Subtract 32 from each side.**

$35 = C$ **Multiply by $\frac{5}{9}$, the reciprocal of $\frac{9}{5}$.**

If you forget the
meaning of a
vocabulary word,
look in the Glossary
or Index.

A temperature of 35°C or lower may indicate hypothermia.

STUDY TIP

Calculate All Steps

If you get confused when calculations are
done mentally, just write the steps on a
sheet of paper.

STUDY TIP

Using Cross-References

If you get stuck on an example, look at
Student Help notes, the Study Guide, or the
Glossary.

Questions

1. In Example 6, why was 95 substituted for F in the formula?

2. What is the purpose of Example 6? How could you use this example if you
   were solving a problem that gave you the Celsius temperature and asked you
   to find the Fahrenheit temperature?

3. Apply the process shown in Example 6 to find the base length b of a triangle
   with a height h of 5 inches and area A of 20 square inches. Use the formula
   $A = \frac{1}{2}bh$. Show your work.

Visual Glossary

The Study Guide on page 130 lists the key vocabulary for Chapter 3. Use the
visual glossary below to help you understand some of the key vocabulary in
Chapter 3. You may want to copy these diagrams into your notebook and refer
to them as you complete the chapter.

GLOSSARY

equivalent equations
(p. 132) Equations with the
same solution(s).

inverse operations (p. 133)
Two operations that undo
each other, such as addition
and subtraction.

rate of *a* per *b* (p. 177)
The relationship $\frac{a}{b}$ of two
quantities *a* and *b* that are
measured in different units.

unit rate (p. 177) A rate per
one given unit.

linear equation (p. 134) An
equation in which the variable
is raised to the first power
and does not occur in a
denominator, inside a square
root symbol, or inside an
absolute value symbol.

Representing Equivalent Equations

When you solve an equation, you write an equation equivalent to the
original equation. You can show the equations are equivalent by
modeling with tiles or by substituting the value of the variable in the
original equation.

$$x + 2 = 5$$
$$x + 2 - 2 = 5 - 2$$
$$x = 3$$

inverse operations

$x + 2 = 5$ and $x = 3$ are equivalent equations.

Rates and Unit Rates

Given a rate, you can find the unit rate by writing an equivalent
fraction with a denominator of 1.

$$\frac{165 \text{ miles}}{3 \text{ hours}} = \frac{(165 \div 3) \text{ miles}}{(3 \div 3) \text{ hours}} = \frac{55 \text{ miles}}{1 \text{ hour}}$$

└ rate   └ unit rate

Linear Equations in One Variable

To check whether an equation in one variable *x* is linear, ask the
questions in the table. If all the answers are as indicated, the equation
is linear.

Is *x* raised to the first power?	Yes
Is *x* in a denominator?	No
Is *x* inside a square root symbol?	No
Is *x* inside an absolute value symbol?	No

The equation $7x + 5 = 33$ is linear. These equations are *not* linear.

$$x^3 - 1 = 7 \qquad \frac{3}{x} - 2 = 4 \qquad \sqrt{2x - 1} = 3 \qquad 4 = |x - 2|$$

TEACHER'S NAME _____ CLASS _____ ROOM _____ DATE _____

Lesson Plan

2-day lesson (See *Pacing the Chapter,* TE page 128A)   For use with pages 131–137

GOAL Solve linear equations using addition and subtraction.

State/Local Objectives _____

_____

✓ **Check the items you wish to use for this lesson.**

STARTING OPTIONS

_____ Prerequisite Skills Review: CRB pages 5–6
_____ Strategies for Reading Mathematics: CRB pages 7–8
_____ Warm-Up: CRB page 11 or Transparencies

TEACHING OPTIONS

_____ Developing Concepts: SE page 131; CRB page 12 (Activity Support Master)
_____ Lesson Opener: CRB page 13 or Transparencies
_____ Examples Day 1: 1–2, SE pages 133; Day 2: 3, SE page 134
_____ Extra Examples: TE pages 133–134 or Transparencies; Internet Help at *www.mcdougallittell.com*
_____ Checkpoint Exercises: Day 1: Exs. 1–6, SE page 133; Day 2: Exs. 7–9, SE page 134
_____ Concept Check: TE page 134
_____ Guided Practice Exercises: SE page 135; Day 1: Exs. 1–15; Day 2: Exs. 16–18

APPLY/HOMEWORK

Homework Assignment

_____ Transitional: Day 1: pp. 135–137 Exs. 19–21, 25–30, 43–45, 65–81 odd;
      Day 2: pp. 136–137 Exs. 52, 54–56, 63, 64, 66–82 even
_____ Average: Day 1: pp. 135–137 Exs. 22–24, 31–36, 46–48, 65–71;
      Day 2: pp. 136–137 Exs. 53, 57, 60, 63, 64, 72–78
_____ Advanced: Day 1: pp. 135–137 Exs. 37–42, 49–51, 67, 68, 72–74, 79–82;
      Day 2: pp. 136–137 Exs. 52–64*, 67, 68, 79–82; EC: CRB p. 20

Reteaching the Lesson

_____ Practice Masters: CRB pages 14–15 (Level A, Level B)
_____ Reteaching with Practice: CRB pages 16–17 or Practice Workbook with Examples;
      Resources in Spanish
_____ Personal Student Tutor: CD-ROM

Extending the Lesson

_____ Interdisciplinary/Real-Life Applications: CRB page 19
_____ Challenge: CRB page 20

ASSESSMENT OPTIONS

_____ Daily Quiz (3.1): TE page 137, CRB page 23, or Transparencies
_____ Standardized Test Practice: SE page 137; STP Workbook; Transparencies

Notes _____

_____

_____

TEACHER'S NAME _____ CLASS _____ ROOM _____ DATE _____

Lesson Plan for Block Scheduling

1-block lesson (See *Pacing the Chapter,* TE page 128A)   For use with pages 131–137

GOAL   Solve linear equations using addition and subtraction.

State/Local Objectives _____

_____

_____

CHAPTER PACING GUIDE	
Day	**Lesson**
1	**3.1 (all)**
2	3.2 (all); 3.3 (all)
3	3.4 (all)
4	3.5 (all); 3.6 (begin)
5	3.6 (end); 3.7 (all)
6	3.8 (all); 3.9 (all)
7	Ch. 3 Review and Assess

✓ **Check the items you wish to use for this lesson.**

STARTING OPTIONS
____ Prerequisite Skills Review: CRB pages 5–6
____ Strategies for Reading Mathematics: CRB pages 7–8
____ Warm-Up: CRB page 11 or Transparencies

TEACHING OPTIONS
____ Developing Concepts: SE page 131; CRB page 12 (Activity Support Master)
____ Lesson Opener: CRB page 13 or Transparencies
____ Examples: 1–3, SE pages 133–134
____ Extra Examples: TE pages 133–134 or Transparencies; Internet Help at *www.mcdougallittell.com*
____ Checkpoint Exercises: Exs. 1–6, SE page 133; Exs. 7–9, SE page 134
____ Concept Check: TE page 134
____ Guided Practice Exercises: SE page 135, Exs. 1–18

APPLY/HOMEWORK
Homework Assignment
____ Block Schedule: pp. 135–137 Exs. 22–24, 31–36, 46–48, 53, 57, 60, 65–78

Reteaching the Lesson
____ Practice Masters: CRB pages 14–15 (Level A, Level B)
____ Reteaching with Practice: CRB pages 16–17 or Practice Workbook with Examples; Resources in Spanish
____ Personal Student Tutor: CD-ROM

Extending the Lesson
____ Interdisciplinary/Real-Life Applications: CRB page 19
____ Challenge: CRB page 20

ASSESSMENT OPTIONS
____ Daily Quiz (3.1): TE page 137, CRB page 23, or Transparencies
____ Standardized Test Practice: SE page 137; STP Workbook; Transparencies

Notes _____

_____

_____

NAME _____ DATE _____

WARM-UP EXERCISES

For use before Lesson 3.1, pages 131–137

Add or subtract.

1. $7 - 12$

2. $-3 - 4$

3. $-6 + 2$

4. $9 - (-1)$

5. $-12 - (-18)$

··

DAILY HOMEWORK QUIZ

For use after Lesson 2.8, pages 113–118

Find the quotient.

1. $-35 \div 3\frac{1}{2}$

2. $\dfrac{11}{\frac{1}{3}}$

3. Simplify $\dfrac{25d - 125}{5}$.

4. Evaluate $\dfrac{3x - 8}{y}$ when $x = 2$ and $y = 24$.

5. What is the domain of the function $y = \dfrac{3x}{x - 3}$?

Lesson 3.1

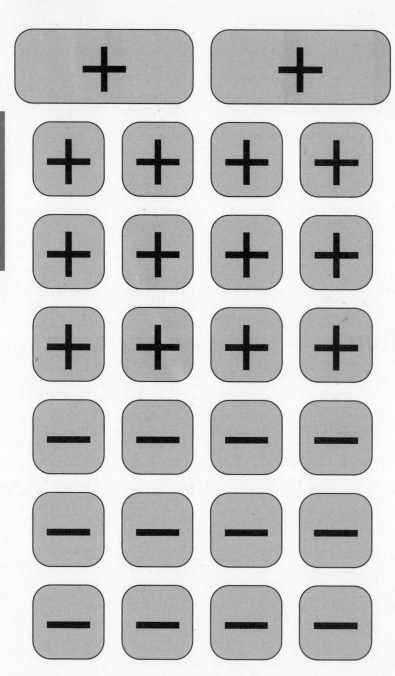

NAME _____ DATE _____

Activity Lesson Opener

For use with pages 131–137

SET UP: Work in a group.

1. Identify the set of equations that has been assigned to your group.

Set 1	*Set 2*
$n + 5 = -3$	$t - 3 = 4$
$6 = r - 8$	$7 + w = -5$
$-2 + y = -5$	$6 = x + 4$
$p - 4 = 5$	$r - 6 = -8$
$2 = m + 3$	$-6 = y - 2$
Set 3	*Set 4*
$2 = x - 5$	$-2 = r + 4$
$a + 6 = -2$	$5 + z = 3$
$-3 + d = -4$	$-8 + x = -3$
$m - 2 = 7$	$m - 9 = 6$
$-7 = z + 2$	$4 = 8 + n$

2. As your teacher calls out one of the steps shown at the right, check each equation in your list to see if it can be solved using that step. If it can, use the step to solve the equation and write the solution next to the equation. The first group to solve all of their equations correctly wins!

Steps to use on an equation

Add 2 to each side.
Add 3 to each side.
Add 4 to each side.
Add 5 to each side.
Add 6 to each side.
Add 8 to each side.
Add 9 to each side.
Subtract 2 from each side.
Subtract 3 from each side.
Subtract 4 from each side.
Subtract 5 from each side.
Subtract 6 from each side.
Subtract 7 from each side.
Subtract 8 from each side.

Practice A

For use with pages 131–137

State the inverse operation.

1. Add 12.

2. Add -6.

3. Subtract 21.

4. Subtract -15.

5. Add -48.

6. Subtract -30.2.

Check whether the given number is a solution of the equation.

7. $x - 4 = 8; 12$

8. $x + 9 = 5; 4$

9. $x - 12 = 9; 3$

10. $18 = x - 7; 25$

11. $13 + x = 9; -4$

12. $22 = 14 + x; 8$

Solve the equation.

13. $x = 8 - 12$

14. $x + 2 = 14$

15. $x - 3 = 6$

16. $x - 6 = 15$

17. $x + 17 = 20$

18. $28 = 17 + x$

19. $-1 = x - 4$

20. $x + 8 = 7$

21. $13 = 6 + x$

22. $x + 4 = -7$

23. $1 = 5 + x$

24. $3 = -12 + x$

25. $x - 3 = -10$

26. $-10 = -6 + x$

27. $-5 + x = -2$

Simplify first and then solve the equation.

28. $x - (-6) = -4$

29. $-7 + x = 13$

30. $-12 = -9 + x$

31. $6 - 2 - x = 10$

32. $x + 9 + 15 = 30 - 12$

33. $-x - (-10) = 5$

In Exercises 34–37, write and solve an equation to answer the question.

34. *Altitude* An airplane was at a cruising altitude, then descended 3000 feet. If the airplane is at 21,000 feet now, what was the cruising altitude?

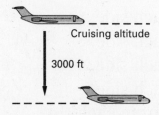

35. *Mountains* Mt. Whitney is the highest point in the contiguous United States at 14,494 feet. This is 5826 feet less than Mt. McKinley in Alaska. How high is Mt. McKinley?

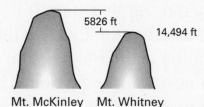

36. *Hiking* You hiked 8.3 miles in Denali National Park which is 3 miles farther than you hiked yesterday. How far did you hike yesterday?

37. *Gold* You spent 28 days panning for gold. This was 17 days less than last year. How many days did you pan for gold last year?

NAME _____ DATE _____

Practice B
For use with pages 131–137

State the inverse operation.

1. Add 18.   **2.** Add -11.   **3.** Subtract 31.

4. Subtract -25.   **5.** Subtract $-4\frac{1}{2}$.   **6.** Subtract -10.06.

Solve the equation.

7. $x = 8 - 17$ **8.** $x + 9 = 14$ **9.** $x - 3 = 11$

10. $-5 = x - 4$ **11.** $x + 10 = 7$ **12.** $x - 15 = 6$

13. $x + 14 = -7$ **14.** $-12 = 12 + x$ **15.** $-15 = -13 + x$

16. $-\frac{3}{4} = x + \frac{1}{2}$ **17.** $-\frac{1}{3} + x = \frac{2}{3}$ **18.** $3\frac{1}{4} + x = 4\frac{3}{4}$

19. $x - 2\frac{5}{6} = -2\frac{1}{2}$ **20.** $12.6 = x - 7.5$ **21.** $8.4 + x = -3.9$

22. $x - (-9) = -9$ **23.** $-x + (-3) = 13$ **24.** $-11 = -2 + x$

25. $10 = -x - 8 - 6$ **26.** $15 - (-x) = 23$ **27.** $-x - (-10) = -24$

Find the length of the side marked x.

28. The perimeter is 30 feet.

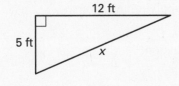

29. The perimeter is 44 centimeters.

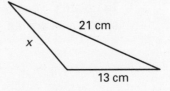

In Exercises 30–33, write and solve an equation to answer the question.

30. *Altitude* An airplane was at a cruising altitude, then descended 3800 feet. If the airplane is at 28,000 feet now, what was the cruising altitude?

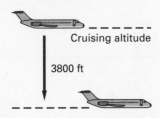

31. *Mountains* Mt. McKinley is the highest point in the United States at 20,320 feet. This is 8708 feet less than Mt. Everest. How high is Mt. Everest?

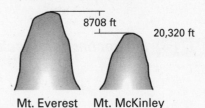

32. *Hiking* You hiked 8.3 miles in Denali National Park which is 2.7 miles farther than you hiked yesterday. How far did you hike yesterday?

33. *Temperature* To convert a Celsius temperature, C, to a Kelvin temperature, K, add 273.15 to the Celsius temperature. What is $-16°C$ in degrees Kelvin?

NAME _____   DATE _____

Reteaching with Practice

For use with pages 131–137

GOAL   Solve linear equations using addition and subtraction

VOCABULARY

Equivalent equations have the same solutions.

Inverse operations are two operations that undo each other, such as addition and subtraction.

Each time you apply a transformation to an equation, you are writing a **solution step.**

In a **linear equation,** the variable is raised to the *first* power and does not occur inside a square root symbol, an absolute value symbol, or in a denominator.

EXAMPLE 1   *Adding to Each Side*

Solve $y - 7 = -2$.

SOLUTION

To isolate y, you need to undo the subtraction by applying the inverse operation of adding 7.

$$y - 7 = -2 \qquad \text{Write original equation.}$$
$$y - 7 + 7 = -2 + 7 \qquad \text{Add 7 to each side.}$$
$$y = 5 \qquad \text{Simplify.}$$

The solution is 5. Check by substituting 5 for y in the original equation.

Exercises for Example 1

Solve the equation.

1. $t - 11 = 4$ **2.** $x - 2 = -3$ **3.** $5 = d - 8$

EXAMPLE 2   *Subtracting from Each Side*

Solve $q + 4 = -9$.

SOLUTION

To isolate q, you need to undo the addition by applying the inverse operation of subtracting 4.

$$q + 4 = -9 \qquad \text{Write original equation.}$$
$$q + 4 - 4 = -9 - 4 \qquad \text{Subtract 4 from each side.}$$
$$q = -13 \qquad \text{Simplify.}$$

The solution is -13. Check by substituting -13 for q in the original equation.

Algebra 1
Chapter 3 Resource Book

Reteaching with Practice

For use with pages 131–137

Exercises for Example 2

Solve the equation.

4. $s + 1 = -8$ **5.** $-6 + b = 10$ **6.** $6 = w + 12$

EXAMPLE 3 *Simplifying First*

Solve $x - (-3) = 10$.

SOLUTION

$$x - (-3) = 10 \qquad \text{Write original equation.}$$
$$x + 3 = 10 \qquad \text{Simplify.}$$
$$x + 3 - 3 = 10 - 3 \qquad \text{Subtract 3 from each side.}$$
$$x = 7 \qquad \text{Simplify.}$$

The solution is 7. Check by substituting 7 for x in the original equation.

Exercises for Example 3

Solve the equation.

7. $8 + z = 1$ **8.** $7 = k - 2$ **9.** $9 = a + (-5)$

EXAMPLE 4 *Modeling a Real-Life Problem*

The original price of a bicycle was marked down $20 to a sale price of $85. What was the original price?

SOLUTION

Original price (p) − Price reduction (20) = Sale Price (85)

Solve the equation $p - 20 = 85$.

$$p - 20 = 85 \qquad \text{Write real-life equation.}$$
$$p - 20 + 20 = 85 + 20 \qquad \text{Add 20 to each side.}$$
$$p = 105 \qquad \text{Simplify.}$$

The original price was $105. Check this in the statement of the problem.

Exercise for Example 4

10. After a sale, the price of a stereo was marked up $35 to a regular price of $310. What was the sale price?

NAME _____ DATE _____

Quick Catch-Up for Absent Students

For use with pages 131–137

The items checked below were covered in class on (date missed) _____

Developing Concepts Activity: One-Step Equations (p. 131)

____ **Goal:** Use algebra tiles to solve one-step equations.

Lesson 3.1: Solving Equations Using Addition and Subtraction (p. 132–134)

____ **Goal:** Solve linear equations using addition and subtraction.

Material Covered:

____ Student Help: Study Tip

____ Example 1: Add to Each Side of an Equation

____ Student Help: Study Tip

____ Example 2: Simplify First

____ Example 3: Model Temperature Change

Vocabulary:

equivalent equations, p. 132   linear equation, p. 134

inverse operations, p. 133

____ Other (specify) _____

_____

Homework and Additional Learning Support

____ Textbook (specify) pp. 135–137 _____

_____

____ Internet: Extra examples at www.mcdougallittell.com

____ *Reteaching with Practice* worksheet (specify exercises)_____

____ *Personal Student Tutor* for Lesson 3.1

NAME _____ DATE _____

Real-Life Application:
When Will I Ever Use This?

For use with pages 131–137

College Football Stadiums

College football has become a fall tradition around the country. Its popularity can be seen through the stadiums themselves. Increasing attendance at these football games, often over one hundred thousand people, has warranted an almost constant state of expansion. The five biggest college football stadiums have jostled positions on the list numerous times as addition after addition is built to increase seating capacity. The largest stadiums have not been able to go more than ten years without adding on. Desperate for seating, one school cut its stadium into sections, raised it eight feet by hydraulic jacks, and inserted pre-cast seating forms within its inner circle. Some schools have even made "seating adjustments," making bleacher seats narrower to accommodate more people.

The following table shows five college football stadiums by seating capacity.

University	Stadium	Seating capacity
University of Michigan	Michigan Stadium	107,501
Ohio State University	Ohio Stadium	103,801
University of Tennessee	Neyland Stadium	102,854
UCLA	Rose Bowl	98,636
Penn State University	Beaver Stadium	93,967

In Exercises 1–4, write and solve an addition or subtraction equation to answer the question.

1. How many seats would have to be added to Beaver Stadium for it to have as many seats as Michigan Stadium?

2. How many seats would have to be added to Neyland Stadium to make it the largest college football stadium by seating capacity?

3. Many lists do not include the Rose Bowl because it is not actually on a college campus. With that consideration, the University of Georgia's Sanford Stadium holds the fifth spot. Sanford Stadium has 7850 fewer seats than Beaver Stadium. What is the seating capacity of Sanford Stadium?

4. Marcana Municipal Stadium in Rio de Janeiro, Brazil can seat 47,499 more people than Michigan Stadium. What is the seating capacity of Marcana Municipal Stadium?

Challenge: Skills and Applications

For use with pages 131–137

For Exercises 1–8, solve the equation.

1. $4 + |a| = 6$

2. $|m| - 7 = 4$

3. $|t| - 6 = -\frac{1}{2}$

4. $5 - |r| = -3$

5. $-(|b| + 2) = -15$

6. $-(|x| - 5) = -2$

7. $-(|p| - 10) = 4$

8. $-(3 - |s|) = 20$

For Exercises 9–10, use the following information.

Marisa Black's pasture has four straight sides. Three sides are each $\frac{3}{8}$ miles long. It took $1\frac{3}{8}$ miles of fencing to enclose the pasture. Marisa wants to know the length of the fourth side.

9. Use a verbal model to write an equation for this situation, where x is the length of the fourth side.

10. Solve the equation from Exercise 9 and interpret the solution.

For Exercises 11–13, use the following information.

Anne is $2\frac{1}{2}$ inches taller than Taro. There is a $4\frac{1}{4}$-inch difference between Taro's height and Peter's height. Jamala is taller than Anne, but $\frac{3}{4}$ inch shorter than Peter.

11. Order the four people from shortest to tallest.

12. Let x represent Taro's height. Write and simplify an expression involving x for Jamala's height.

13. Jamala's height is 58 inches. Use your expression from Exercise 12 to write an equation. Solve your equation to find Taro's height.

Algebra 1
Chapter 3 Resource Book

TEACHER'S NAME _____ CLASS _____ ROOM _____ DATE _____

Lesson Plan

1-day lesson (See *Pacing the Chapter,* TE page 128A)　　　　　　For use with pages 138–143

GOAL　**Solve linear equations using multiplication and division.**

State/Local Objectives _____

_____

✓ **Check the items you wish to use for this lesson.**

STARTING OPTIONS
____ Homework Check (3.1): TE page 135; Answer Transparencies
____ Homework Quiz (3.1): TE page 137, CRB page 23, or Transparencies
____ Warm-Up: CRB page 23 or Transparencies

TEACHING OPTIONS
____ Lesson Opener: CRB page 4 or Transparencies
____ Examples 1–4: SE pages 138–140
____ Extra Examples: TE pages 139–140 or Transparencies; Internet Help at *www.mcdougallittell.com*
____ Checkpoint Exercises: SE pages 139–140
____ Concept Check: TE page 140
____ Guided Practice Exercises: SE page 141

APPLY/HOMEWORK
Homework Assignment
____ Transitional: pp. 141–143, Exs. 16–18, 31, 32, 37–39, 46–48, 51, 54–56, 57–77 odd
____ Average: pp. 141–143, Exs. 19–21, 25–27, 33, 34, 40–42, 49, 50, 54–59, 63–65, 73–75
____ Advanced: pp. 141–143, Exs. 28–30, 35, 36, 43–45, 51–56, 60–62, 66–69, 76–78; EC: CRB p. 31

Reteaching the Lesson
____ Practice Masters: CRB pages 25–26 (Level A, Level B)
____ Reteaching with Practice: CRB pages 27–28 or Practice Workbook with Examples;
　　　Resources in Spanish
____ Personal Student Tutor: CD-ROM

Extending the Lesson
____ Interdisciplinary/Real-Life Applications: CRB page 30
____ Challenge: CRB page 31

ASSESSMENT OPTIONS
____ Daily Quiz (3.2): TE page 143, CRB page 34, or Transparencies
____ Standardized Test Practice: SE page 143; STP Workbook; Transparencies

Notes _____

_____

_____

Lesson Plan for Block Scheduling

Half-block lesson (See *Pacing the Chapter*, TE page 128A)   **For use with pages 138–143**

GOAL   Solve linear equations using multiplication and division.

State/Local Objectives _____

_____

_____

✓ **Check the items you wish to use for this lesson.**

STARTING OPTIONS

____ Homework Check (3.1): TE page 135; Answer Transparencies

____ Homework Quiz (3.1): TE page 137,
   CRB page 23, or Transparencies

____ Warm-Up: CRB page 23 or Transparencies

TEACHING OPTIONS

____ Lesson Opener: CRB page 4 or Transparencies

____ Examples 1–4: SE pages 138–140

____ Extra Examples: TE pages 139–140 or Transparencies; Internet Help at *www.mcdougallittell.com*

____ Checkpoint Exercises: SE pages 139–140

____ Concept Check: TE page 140

____ Guided Practice Exercises: SE page 141

APPLY/HOMEWORK

Homework Assignment (See also the assignment for Lesson 3.3)

____ Block Schedule: pp. 141–143 Exs. 19–21, 25–27, 33, 34, 40–42, 49, 50, 54–59, 63–65, 73–75

Reteaching the Lesson

____ Practice Masters: CRB pages 25–26 (Level A, Level B)

____ Reteaching with Practice: CRB pages 27–28 or Practice Workbook with Examples;
   Resources in Spanish

____ Personal Student Tutor: CD-ROM

Extending the Lesson

____ Interdisciplinary/Real-Life Applications: CRB page 30

____ Challenge: CRB page 31

ASSESSMENT OPTIONS

____ Daily Quiz (3.2): TE page 143, CRB page 34, or Transparencies

____ Standardized Test Practice: SE page 143; STP Workbook; Transparencies

Notes _____

_____

_____

CHAPTER PACING GUIDE	
Day	**Lesson**
1	3.1 (all)
2	**3.2 (all);** 3.3(all)
3	3.4 (all)
4	3.5 (all); 3.6 (begin)
5	3.6 (end); 3.7 (all)
6	3.8 (all); 3.9 (all)
7	Ch. 3 Review and Assess

LESSON
3.2
NAME _____ DATE _____

WARM-UP EXERCISES

For use before Lesson 3.2, pages 138–143

Multiply or divide.

1. $\left(\dfrac{4}{5}\right)\left(\dfrac{5}{4}\right)$

2. $\dfrac{1}{2} \cdot 2$

3. $\dfrac{21}{3} \cdot 3$

4. $\dfrac{2}{3} \div \dfrac{1}{3}$

5. $\left(-\dfrac{5}{9}\right)\left(-\dfrac{9}{5}\right)$

DAILY HOMEWORK QUIZ

For use after Lesson 3.1, pages 131–137

Solve the equation.

1. $x - 6 = -13$

2. $n - (-4) = 9$

3. $\dfrac{1}{5} + t = \dfrac{3}{5}$

4. $-12 = r - 20$

5. A man walked from a valley, at an elevation of 64 feet below sea level, to the top of a hill 53 feet above sea level. By how many feet did he increase his elevation?

Lesson 3.2

Activity Lesson Opener

For use with pages 138–143

SET UP: Work with a partner. You will need: algebra tiles

1. You can use algebra tiles to solve an equation like $3x = 6$. Use your algebra tiles to model the steps given below.

   a. Place three *x*-tiles on the left and six 1-tiles on the right.

   b. To find the value of *x*, split the tiles on each side of the equation in thirds to get

   $x =$ _____.

In Questions 2–4, use algebra tiles to model and solve the equation. The model for the first equation is given.

2. $2x = -6$

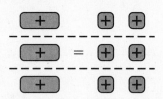

3. $4x = 20$ 4. $3x = -12$

5. Use the equations you have solved to make a conjecture about solving a multiplication equation.

NAME _____ DATE _____

Practice A

For use with pages 138–143

State the inverse operation.

1. Divide by 7.

2. Divide by -4.

3. Multiply by 3.

4. Multiply by -6.

5. Divide by $-\dfrac{1}{3}$.

6. Multiply by $-\dfrac{3}{5}$.

Complete the sentence.

7. To isolate the variable in $\dfrac{1}{4}x$, multiply by _____ or divide by _____.

8. To isolate the variable in $-\dfrac{3}{5}x$, multiply by _____ or divide by _____.

9. To isolate the variable in $\dfrac{x}{2}$, multiply by _____ or divide by _____.

Tell whether the equations are equivalent.

10. $3x = 24$ and $x = 8$

11. $-12x = -3$ and $x = 4$

12. $\dfrac{x}{6} = 30$ and $x = 5$

13. $\dfrac{2}{3}x = -8$ and $x = -12$

Solve the equation.

14. $4x = 12$

15. $-5x = 40$

16. $-32 = 16x$

17. $-3x = -18$

18. $5 = \dfrac{1}{4}x$

19. $7 = 35x$

20. $-6 = \dfrac{3}{4}x$

21. $3 = -9x$

22. $-x = \dfrac{1}{2}$

23. $-\dfrac{1}{2}x = -6$

24. $\dfrac{x}{-4} = 5$

25. $-15 = \dfrac{x}{-2}$

26. $\dfrac{x}{3} = \dfrac{1}{6}$

27. $\dfrac{x}{-3} = 1$

28. $-\dfrac{2}{3}x = \dfrac{4}{9}$

In Exercises 29 and 30, write and solve an equation to answer the question.

29. *Dimensions of a Banner* You are working on a banner for Friday's pep rally. The length of the banner is 4 times the width. The length is 12 feet. What is the width?

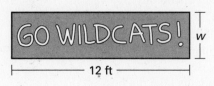

30. *Running Laps* In gym class you run $1\frac{1}{2}$ miles on the track. One lap is $\frac{1}{4}$ mile. How many laps do you run?

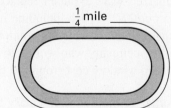

Lesson 3.2

NAME _____ DATE _____

Practice B

For use with pages 138–143

State the inverse operation.

1. Divide by 8.

2. Divide by -5.

3. Multiply by 10.

4. Multiply by $\frac{1}{2}$.

5. Divide by $-\frac{2}{3}$.

6. Multiply by $-\frac{6}{5}$.

Complete the sentence.

7. To isolate the variable in $\frac{1}{8}x$, multiply by _____ or divide by _____.

8. To isolate the variable in $-\frac{7}{10}x$, multiply by _____ or divide by _____.

9. To isolate the variable in $\frac{x}{-2}$, multiply by _____ or divide by _____.

Solve the equation.

10. $4x = 52$

11. $-5x = 80$

12. $-96 = 16x$

13. $-3x = -27$

14. $\frac{1}{3} = -x$

15. $9 = 45x$

16. $-6 = -9x$

17. $3 = \frac{x}{11}$

18. $\frac{x}{-4} = 17$

19. $-15 = \frac{x}{-8}$

20. $-\frac{1}{2}x = -30$

21. $-12 = \frac{3}{4}x$

22. $\frac{x}{-3} = \frac{5}{6}$

23. $-\frac{2}{3}x = \frac{4}{15}$

24. $-\frac{3}{4}x = -6\frac{1}{2}$

25. $0 = \frac{2}{3}x$

26. $\frac{4}{3}x = 32$

27. $-4x = -18$

In Exercises 28–31, write and solve an equation to answer the question.

28. *Sports and Drama* About $\frac{2}{5}$ of students participate in sports and drama. In what size school would you expect to find 130 students in sports and drama?

29. *Posters* You purchased 8 copies of a poster and paid $78. How much did each poster cost?

30. The Schatzels want to put 11 replacement windows in their house. The windows cost $234 each including installation. They can't spend any more than $2500. How many windows can they replace? If they looked for another company who would do all 11 windows for $2500, how much would each window cost (round to the nearest whole dollar)?

31. At George Washington Elementary School, $\frac{5}{8}$ of the students take the bus. If 330 students take the bus, how many students attend the school?

NAME _____ DATE _____

Reteaching with Practice

For use with pages 138–143

GOAL Solve linear equations using multiplication and division.

> **VOCABULARY**
>
> **Properties of equality** are rules of algebra that can be used to transform equations into equivalent equations.

EXAMPLE 1 *Dividing Each Side of an Equation*

Solve $7n = -35$.

SOLUTION

To isolate n, you need to undo the multiplication by applying the inverse operation of dividing by 7.

$$7n = -35 \qquad \text{Write original equation.}$$

$$\frac{7n}{7} = \frac{-35}{7} \qquad \text{Divide each side by 7.}$$

$$n = -5 \qquad \text{Simplify.}$$

The solution is -5. Check by substituting -5 for n in the original equation.

Exercises for Example 1

Solve the equation.

1. $-12x = 6$ **2.** $4 = 24y$ **3.** $-5z = -35$

EXAMPLE 2 *Multiplying Each Side of an Equation*

Solve $-\dfrac{3}{4}t = 9$.

SOLUTION

To isolate t, you need to multiply by the reciprocal of the fraction.

$$-\frac{3}{4}t = 9 \qquad \text{Write original equation.}$$

$$\left(-\frac{4}{3}\right)\left(-\frac{3}{4}\right)t = \left(-\frac{4}{3}\right)9 \qquad \text{Multiply each side by } -\frac{4}{3}.$$

$$t = -12 \qquad \text{Simplify.}$$

The solution is -12. Check by substituting -12 for t in the original equation.

Reteaching with Practice

Exercises for Example 2

Solve the equation.

4. $\frac{1}{6}c = -2$　　　　**5.** $\frac{f}{7} = 3$　　　　**6.** $\frac{2}{3}q = 12$

EXAMPLE 3 *Modeling a Real-Life Problem*

Write and solve an equation to find your average speed s on a plane flight. You flew 525 miles in 1.75 hours.

SOLUTION

Verbal Model

$$\boxed{\begin{array}{c}\text{Speed}\\\text{of jet}\end{array}} \cdot \boxed{\text{Time}} = \boxed{\text{Distance}}$$

Labels　　Speed of jet $= s$　　　　(miles per hour)
　　　　　　　Time $= 1.75$　　　　　　　　(hours)
　　　　　　　Distance $= 525$　　　　　　　(miles)

Algebraic Model

$$s(1.75) = 525 \qquad \text{Write algebraic model.}$$

$$\frac{s(1.75)}{1.75} = \frac{525}{1.75} \qquad \text{Divide each side by 1.75.}$$

$$s = 300 \qquad \text{Simplify.}$$

The speed s was 300 miles per hour. Check this in the statement of the problem.

Exercises for Example 3

7. Write and solve an equation to find your average speed in an airplane if you flew 800 miles in 2.5 hours.

8. Write and solve an equation to find your time in an airplane if you flew 1530 miles at a speed of 340 miles per hour.

NAME _____ DATE _____

Quick Catch-Up for Absent Students

For use with pages 138–143

The items checked below were covered in class on (date missed) _____

Lesson 3.2: Solving Equations Using Multiplication and Division

_____ **Goal:** Solve linear equations using multiplication and division.

Material Covered:

_____ Student Help: Study Tip

_____ Example 1: Divide Each Side of an Equation

_____ Example 2: Multiply Each Side of an Equation

_____ Example 3: Multiply Each Side by a Reciprocal

_____ Example 4: Model a Real-Life Problem

Vocabulary:

properties of equality, p. 140

_____ Other (specify)_____

_____

Homework and Additional Learning Support

_____ Textbook (specify) pp. 141–143 _____

_____

_____ Internet: Extra examples at www.mcdougallittell.com

_____ *Reteaching with Practice* worksheet (specify exercises) _____

_____ *Personal Student Tutor* for Lesson 3.2

NAME _____ DATE _____

Interdisciplinary Application

For use with pages 138–143

Pony Express

HISTORY   In the 1850s, the only way the merchants and bankers on the West Coast could receive information from the East Coast was by ship or by stage-coach, which could take months. To meet the need for transcontinental communication, William Russell, William Waddell, and Alexander Majors founded the Pony Express.

On April 3, 1860, the Pony Express completed the 1966-mile journey from St. Joseph, Missouri to Sacramento, California in just 10 days, which became the standard delivery time. Eighty riders were hired at a wage of $100 per month (4 weeks) to race between 160 stations along the trail that goes through the present day states of Missouri, Kansas, Nebraska, Colorado, Wyoming, Utah, Nevada, and California. Riders switched horses approximately every 10 miles and averaged 75 miles per run. The riders continued day and night no matter the weather, establishing the first high-speed link between the two coasts. Financially, the Pony Express was far from a success. Although the delivery charge was around $5 per half ounce, it did not come close to covering the actual expenses.

In 1861, on the brink of the Civil War, President Lincoln's Inaugural Address was telegraphed from Washington to St. Joseph and then delivered to Sacramento by the Pony Express. This was the fastest trip ever made by the Pony Express, for the riders covered about 255 miles per day.

On October 24, 1861, the transcontinental telegraph was finally completed and the Pony Express became obsolete overnight. Even though it existed for only 18 short months, the Pony Express will always be a legend of the American West.

In Exercises 1-4, use the information above to write and solve a multiplication or division equation to answer the question.

1. How many miles per day did the riders usually travel on the journey from St. Joseph to Sacramento?

2. How long did it take the riders to make the delivery of Lincoln's Inaugural Address?

3. How much did it cost to send a 9-ounce package with the Pony Express?

4. How much would a rider earn delivering for the Pony Express for 30 weeks?

NAME _____ DATE _____

Challenge: Skills and Applications

For use with pages 138–143

For Exercises 1–4, solve the equation.

1. $2|p| = 13$

2. $-12 = -5|v|$

3. $\dfrac{|x|}{-5} = -7$

4. $\dfrac{3}{5}|y| = 42$

For Exercises 5–8, solve the equation.

Example: $\dfrac{3}{a} = 4$

Solution:
$$\dfrac{3}{a} = 4$$
$$a\left(\dfrac{3}{a}\right) = a(4)$$
$$3 = 4a$$
$$\dfrac{3}{4} = a$$

5. $7 = \dfrac{4}{n}$

6. $-\dfrac{1}{2k} = 5$

7. $-3 = \dfrac{5}{b}$

8. $-\dfrac{4}{3x} = 8$

9. What decimal is the reciprocal of 0.4?

10. You can transform an equation into an equivalent equation by multiplying both sides by the same *nonzero* number. Explain why $4x = 12$ is not transformed to an equivalent equation when you multiply both sides by zero.

11. Explain why you cannot transform an equation by dividing both sides by zero.

12. Consuela Lopez drove 20 miles to visit a friend, averaging 25 miles per hour, and then she took a different route back, averaging the same speed. Her entire trip took 2 hours. Write and solve an equation to find the distance d that she drove on the way back.

TEACHER'S NAME _____ CLASS _____ ROOM _____ DATE _____

Lesson Plan

1-day lesson (See *Pacing the Chapter,* TE page 128A)   For use with pages 144–149

GOAL   Use two or more steps to solve a linear equation.

State/Local Objectives _____

_____

✓ **Check the items you wish to use for this lesson.**

STARTING OPTIONS

____ Homework Check (3.2): TE page 141; Answer Transparencies

____ Homework Quiz (3.2): TE page 143, CRB page 34, or Transparencies

____ Warm-Up: CRB page 34 or Transparencies

TEACHING OPTIONS

____ Lesson Opener: CRB page 35 or Transparencies

____ Investigation 3: Algebra Tiles Investigations

____ Examples 1–5: SE pages 144–146

____ Extra Examples: TE pages 145–146 or Transparencies; Internet Help at *www.mcdougallittell.com*

____ Checkpoint Exercises: SE pages 144–146

____ Graphing Calculator Activity with Keystrokes: CRB pages 36–37

____ Concept Check: TE page 146

____ Guided Practice Exercises: SE page 147

APPLY/HOMEWORK

Homework Assignment

____ Transitional: pp. 147–149, Exs. 19–24, 31–33, 40, 43, 44, 53, 54, 55–73 odd, Quiz 1

____ Average: pp. 147–149, Exs. 25–27, 34–36, 41, 45, 46, 53, 54, 56–74 even, Quiz 1

____ Advanced: pp. 147–149, Exs. 28–30, 37–39, 42, 45–54*, 58–60, 64–66, 71–74, Quiz 1; EC: CRB p. 44

Reteaching the Lesson

____ Practice Masters: CRB pages 38–39 (Level A, Level B)

____ Reteaching with Practice: CRB pages 40–41 or Practice Workbook with Examples; Resources in Spanish

____ Personal Student Tutor: CD-ROM

Extending the Lesson

____ Interdisciplinary/Real-Life Applications: CRB page 43

____ Challenge: CRB page 44

ASSESSMENT OPTIONS

____ Daily Quiz (3.3): TE page 149, CRB page 48, or Transparencies

____ Standardized Test Practice: SE page 148; STP Workbook; Transparencies

____ Quiz 3.1–3.3: SE page 149; CRB page 45; Resources in Spanish

Notes _____

_____

_____

Algebra 1
Chapter 3 Resource Book

TEACHER'S NAME _____ CLASS _____ ROOM _____ DATE _____

Lesson Plan for Block Scheduling

Half-block lesson (See *Pacing the Chapter*, TE page 128A)   **For use with pages 144–149**

GOAL   Use two or more steps to solve a linear equation.

State/Local Objectives _____

_____

_____

✓ **Check the items you wish to use for this lesson.**

STARTING OPTIONS

____ Homework Check (3.2): TE page 141; Answer Transparencies

____ Homework Quiz (3.2): TE page 143,
CRB page 34, or Transparencies

____ Warm-Up: CRB page 34 or Transparencies

TEACHING OPTIONS

____ Lesson Opener: CRB page 35 or Transparencies

____ Investigation 3: Algebra Tiles Investigations

____ Examples 1–5: SE pages 144–146

____ Extra Examples: TE pages 145–146 or Transparencies; Internet Help at *www.mcdougallittell.com*

____ Checkpoint Exercises: SE pages 144–146

____ Graphing Calculator Activity with Keystrokes: CRB pages 36–37

____ Concept Check: TE page 146

____ Guided Practice Exercises: SE page 147

APPLY/HOMEWORK

Homework Assignment (See also assignment for Lesson 3.2.)

____ Block Schedule: pp. 147–149 Exs. 25–27, 34–36, 41, 45, 46, 53, 54, 56–74 even, Quiz 1

Reteaching the Lesson

____ Practice Masters: CRB pages 38–39 (Level A, Level B)

____ Reteaching with Practice: CRB pages 40–41 or Practice Workbook with Examples; Resources in Spanish

____ Personal Student Tutor: CD-ROM

Extending the Lesson

____ Interdisciplinary/Real-Life Applications: CRB page 43

____ Challenge: CRB page 44

ASSESSMENT OPTIONS

____ Daily Quiz (3.3): TE page 149, CRB page 48, or Transparencies

____ Standardized Test Practice: SE page 148; STP Workbook; Transparencies

____ Quiz 3.1–3.3: SE page 149; CRB page 45; Resources in Spanish

Notes _____

_____

_____

CHAPTER PACING GUIDE	
Day	Lesson
1	3.1 (all)
2	3.2 (all); **3.3 (all)**
3	3.4 (all)
4	3.5 (all); 3.6 (begin)
5	3.6 (end); 3.7 (all)
6	3.8 (all); 3.9 (all)
7	Ch. 3 Review and Assess

Lesson 3.3

WARM-UP EXERCISES

For use before Lesson 3.3, pages 144–149

Simplify.

1. $2(3 - 9x)$

2. $3x - 7 - 8x$

3. $-7(2x - 11)$

4. $7x - 5x + 3$

5. $6x - 2(4x - 1)$

DAILY HOMEWORK QUIZ

For use after Lesson 3.2, pages 138–143

Solve the equation.

1. $5x = -35$

2. $-8b = 72$

3. $\dfrac{t}{3} = -2$

4. $\dfrac{y}{-6} = -3$

5. Four friends dining in a restaurant decide to split the bill evenly between them. Each person will pay $9.45. How much is the total bill? Use the verbal model to write an equation to solve the problem.

$$\frac{\boxed{\text{Total Bill}}}{\boxed{\substack{\text{Number} \\ \text{of people}}}} = \boxed{\substack{\text{Cost per} \\ \text{person}}}$$

Algebra 1
Chapter 3 Resource Book

Available as
a transparency

1. An electrician charges $15 per service call plus $25 an hour plus the cost of the parts used. The total bill for a recent job was $155. If the parts for this job cost $65 and x represents the number of hours worked, which equation can be used to model this problem? Why?

   A. $15 + 25 + x + 65 = 155$

   B. $15x + 25 + 65 = 155$

   C. $15 + 25x + 65 = 155$

   D. $15 + 25 + 65x = 155$

2. On a recent trip to a computer store, you purchased one software package for $24.99 and three printer cartridges. The total bill with tax was $90.91. If the tax was $5.95 and x represents the cost of each printer cartridge, which equation can be used to model this problem? Why?

   A. $24.99 + 3x + 5.95 = 90.91$

   B. $24.99 + 3 + x + 5.95 = 90.91$

   C. $24.99 + 3x(5.95) = 90.91$

   D. $5.95(24.99 + 3x) = 90.91$

3. You subscribe to a long-distance calling plan that costs $12.99 a month plus $0.09 per minute for all long-distance calls. Your total long-distance bill last month was $21.54. If x represents the number of long-distance minutes you used last month, which equation can be used to model this problem? Explain.

   A. $12.99 + 0.09 + x = 21.54$

   B. $21.54 - 0.09 - x = 12.99$

   C. $12.99x + 0.09 = 21.54$

   D. $0.09x + 12.99 = 21.54$

Lesson 3.3

NAME _____ DATE _____

Graphing Calculator Activity

For use with pages 144–149

GOAL **To solve multi-step equations using the solve feature of a graphing calculator**

You will learn several ways to solve equations using your graphing calculator. One way is to use the *Solve* feature. This feature is also a great way to check your solutions of equations.

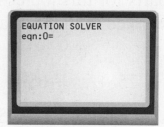

```
EQUATION SOLVER
eqn:0=
```

Activity

❶ To use the *Solve* feature of your graphing calculator it may help to set the equation equal to zero. Rewrite the equation $\frac{3}{4}x - 6 = 9$ so that it is equal to zero by subtracting 9 from each side of the equation.

❷ Enter the equation in the *Solve* feature of your graphing calculator to solve the equation. You may have to select the variable for which to solve.

❸ What is the solution of $\frac{3}{4}x - 15 = 0$?

❹ Repeat Steps 1–3 for each question.

    a. $6x - 3x - 10 = 11$ **b.** $8x - 2(x + 3) = 12$

Exercises

Solve the equation using the *Solve* feature of your graphing calculator.

  1. $3x - 5 = 13$ **2.** $\frac{5}{4}x + 7 = -8$ **3.** $2x + 7 + 8 = 35$

  4. $11x + 3(x - 2) = -34$ **5.** $27 = 9(3 - 2x)$ **6.** $\frac{2}{3}(x + 6) = -12$

See page 37 for keystrokes.

NAME _____ DATE _____

Graphing Calculator Activity Keystrokes

For use with pages 144–149

TI-82

MATH 0

3 X,T,θ ÷ 4 − 15 , X,T,θ , 0)

ENTER

MATH 0

3 X,T,θ − 21 , X,T,θ , 0) ENTER

MATH 0

6 X,T,θ − 18 , X,T,θ , 0) ENTER

TI-83

MATH 0

CLEAR 3 X,T,θ,n ÷ 4 − 15 ENTER

ALPHA [SOLVE]

▲ CLEAR 3 X,T,θ,n − 21 ENTER

ALPHA [SOLVE]

▲ CLEAR 6 X,T,θ,n − 18 ENTER

ALPHA [SOLVE]

SHARP EL-9600c

2ndF [SOLVER]

CL 3 X/θ/T/n ÷ 4 − 15 ENTER

2ndF [EXE]

CL CL CL 3 X/θ/T/n − 21 ENTER

2ndF [EXE]

CL CL CL 6 X/θ/T/n − 18 ENTER

2ndF [EXE]

CASIO CFX-9850GA PLUS

From the main menu, choose EQUA.

3 X,θ,T ÷ 4 − 15 EXE

F6

F1 ▲ 3 X,θ,T − 21 EXE

F6

F1 ▲ 3 X,θ,T − 18 EXE

F6

NAME _____ DATE _____

Practice A

For use with pages 144–149

Check whether the given number is a solution of the equation.

1. $5x - 3 = 28; 5$

2. $6x + 4 = -2; -1$

3. $4(x - 3) = -16; -1$

4. $\frac{x}{4} + 6 = -12; -24$

5. $7 - 2x = 13; -3$

6. $\frac{1}{4}x - 8 = -7; -4$

State the first step in solving the equation.

7. $5x + 9 = 24$

8. $3x - 5 = 22$

9. $42 = 6 + 9x$

10. $2(3x - 4) = 29$

11. $33 = 3x + 8x$

12. $8 - 2x = 10$

Solve the equation.

13. $3x + 8 = 32$

14. $5x - 4 = 21$

15. $2x + 3 = 11$

16. $3x - 1 = 8$

17. $5x - 20 = 5$

18. $2x + 5 = -2$

19. $-4 = \frac{1}{2}x + 3$

20. $\frac{2}{3}x + 11 = 7$

21. $\frac{2}{3}x - \frac{2}{3} = 0$

Solve the equation by simplifying both sides and isolating the variable.

22. $2x + 3x = 5$

23. $10x - 3x = 20 + 1$

24. $2(x - 4) = 2$

25. $\frac{1}{3}(x + 6) = 1$

26. $4 = \frac{2}{3}x + 9 + \frac{1}{3}x$

27. $14 = -2(4x + 5)$

28. $17 = 2(2x + 9)$

29. $5x - 8x = 18$

30. $2(x + 5) + 3x = 0$

In Exercises 31 and 32, write and solve an equation to answer the question.

31. *Piano Keyboard* The keyboard of a piano has seven full octaves with 5 black keys in each octave and one extra black key. There are a total of 88 black and white keys on a piano. How many white keys are on a piano?

Part of a keyboard

32. *Band Fundraiser* Your school band needs to buy new percussion equipment. The equipment will cost $2000. You have collected $800 in previous fundraisers. If you sell sandwiches at $4 each, how many sandwiches will you need to sell to raise the remaining funds?

$$\boxed{\text{Cost per sandwich}} \cdot \boxed{\text{Number of sandwiches sold}} + \boxed{\text{Money already raised}} = \boxed{\text{Cost of equipment}}$$

NAME _____ DATE _____

Practice B

For use with pages 144–149

Check whether the given number is a solution of the equation.

1. $7 - 2x = 15; -4$

2. $4(x - 3) = -24; -3$

3. $5x - 8x + 7 = 22; -5$

4. $\dfrac{x}{8} + 6 = -12; -48$

5. $\dfrac{3}{5}x - 9 = -3; -20$

6. $\dfrac{1}{2}(4x - 8) = 8; -2$

Solve the equation.

7. $3x + 5 = 32$

8. $5x - 14 = 21$

9. $\dfrac{x}{2} - 3 = -4$

10. $\dfrac{x}{-3} + 14 = 8$

11. $\dfrac{3}{4}x - \dfrac{3}{4} = 0$

12. $\dfrac{2x}{3} + 9 = -7$

13. $2x - 3x = 11$

14. $-4x + 9x = -30$

15. $6 = 7x + x$

16. $4x + 3x - 9 = 5$

17. $11x - 3x = 12 - 20$

18. $-2(x - 4) = 2$

19. $\dfrac{3}{4}(x + 2) = -1$

20. $4 = \dfrac{2}{3}x + 9 - \dfrac{1}{3}x$

21. $10 = -\dfrac{2}{3}(4x + 5)$

22. $-8 = \dfrac{1}{3}x + x$

23. $5x + 4 - 8x = 13$

24. $3x + 2(x + 5) = 15$

25. $4x - (2x + 3) = 7$

26. $x - (3x - 9) = -5$

27. $13x - 4(2x - 5) = 15$

In Exercises 28–30, write and solve an equation to answer the question.

28. *Piano Keyboard* The keyboard of a piano has seven full octaves plus two extra white keys and one extra black key. There are 36 black keys on a piano. How many black keys are there in one octave?

Part of a keyboard

29. *Band Fundraiser* Your school band needs to buy new percussion equipment. The equipment will cost $2450. You have collected $812 in previous fundraisers. If you sell sandwiches at $3.50 each, how many sandwiches will you need to sell to raise the remaining funds?

| Cost per sandwich | · | Number of sandwiches sold | + | Money already raised | = | Cost of equipment |

30. *Wrapping a Package* It takes 64 inches of ribbon to make a bow and wrap the ribbon around a box. The bow takes 30 inches of ribbon. The width of the box is 12 inches. What is the height of the box?

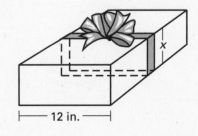

12 in.

Algebra 1
Chapter 3 Resource Book

NAME _____ DATE _____

Reteaching with Practice

For use with pages 144–149

GOAL **Use two or more steps to solve a linear equation**

EXAMPLE 1 *Solving a Linear Equation*

Solve $-3x - 4 = 5$.

SOLUTION

To isolate the variable x, undo the subtraction and then the multiplication.

$-3x - 4 = 5$	Write original equation.
$-3x - 4 + 4 = 5 + 4$	Add 4 to each side.
$-3x = 9$	Simplify.
$\dfrac{-3x}{-3} = \dfrac{9}{-3}$	Divide each side by -3.
$x = -3$	Simplify.

The solution is -3. Check this in the original equation.

Exercises for Example 1

Solve the equation.

1. $5y + 8 = -2$ **2.** $7 - 6m = 1$ **3.** $\dfrac{x}{4} - 1 = 5$

EXAMPLE 2 *Using the Distributive Property and Combining Like Terms*

Solve $y + 5(y + 3) = 33$.

SOLUTION

$y + 5(y + 3) = 33$	Write original equation.
$y + 5y + 15 = 33$	Use distributive property.
$6y + 15 = 33$	Combine like terms.
$6y + 15 - 15 = 33 - 15$	Subtract 15 from each side.
$6y = 18$	Simplify.
$\dfrac{6y}{6} = \dfrac{18}{6}$	Divide each side by 6.
$y = 3$	Simplify.

The solution is 3. Check this in the original equation.

NAME _____ DATE _____

Reteaching with Practice

For use with pages 144–149

Exercises for Example 2

Solve the equation.

4. $4x - 8 + x = 2$

5. $6 - (b + 1) = 9$

6. $10(z - 2) = 1 + 4$

EXAMPLE 3 Solving a Real-Life Problem

The sum of the ages of two sisters is 25. The second sister's age is 5 more than three times the first sister's age n. Find the two ages.

SOLUTION

Verbal Model	First sister's age + Second sister's age = Sum

Labels

First sister's age $= n$ (years)

Second sister's age $= 3n + 5$ (years)

Sum $= 25$ (years)

Algebraic Model

$$n + (3n + 5) = 25 \qquad \text{Write real-life equation.}$$
$$4n + 5 = 25 \qquad \text{Combine like terms.}$$
$$4n + 5 - 5 = 25 - 5 \qquad \text{Subtract 5 from each side.}$$
$$4n = 20 \qquad \text{Simplify.}$$
$$\frac{4n}{4} = \frac{20}{4} \qquad \text{Divide each side by 4.}$$
$$n = 5 \qquad \text{Simplify.}$$

The first sister's age is 5. The second sister's age is $3(5) + 5 = 20$.

Exercises for Example 3

7. A parking garage charges $3 plus $1.50 per hour. You have $12 to spend for parking. Write and solve an equation to find the number of hours that you can park.

8. As a lifeguard, you earn $6 per day plus $2.50 per hour. Write and solve an equation to find how many hours you must work to earn $16 in one day.

NAME _____ DATE _____

Quick Catch-Up for Absent Students

For use with pages 144–149

The items checked below were covered in class on (date missed) _____

Lesson 3.3: Solving Multi-Step Equations

____ **Goal:** Use two or more steps to solve a linear equation.

Material Covered:

____ Student Help: Vocabulary Tip

____ Example 1: Solve a Linear Equation

____ Example 2: Use a Verbal Model

____ Example 3: Combine Like Terms First

____ Student Help: Study Tip

____ Example 4: Use the Distributive Property

____ Student Help: Study Tip

____ Example 5: Multiply by a Reciprocal First

____ Other (specify)_____

_____

Homework and Additional Learning Support

____ Textbook (specify) <u>pp. 147–149</u>_____

_____

____ Internet: Extra examples at www.mcdougallittell.com

____ *Reteaching with Practice* worksheet (specify exercises) _____

____ *Personal Student Tutor* for Lesson 3.3

Algebra 1
Chapter 3 Resource Book

NAME _____ DATE _____

Interdisciplinary Application

For use with pages 144–149

Crickets

BIOLOGY   The male cricket makes a chirping noise by rubbing his front wings together. Although the sound is often used to communicate with and attract female crickets, it has a relationship with air temperature. The higher the air temperature, the greater the number of chirps generated per minute. The following formula, which can approximate the number of chirps per minute given the air temperature or can approximate the air temperature given the number of chirps per minute, is

$$c = 4t - 148,$$

where c is the number of chirps per minute and t is the air temperature in degrees Fahrenheit. For example, you count 80 chirps in one minute.

$80 = 4t - 148$	Substitute 80 for c.
$80 + 148 = 4t - 148 + 148$	Add 148 to each side.
$228 = 4t$	Simplify.
$\dfrac{228}{4} = \dfrac{4t}{4}$	Divide each side by 4.
$t = 57$	Simplify.

The temperature is about 57°F.

1. Find the number of chirps made per minute by a cricket at 65°F.

2. What is the air temperature when a cricket makes 52 chirps per minute?

3. What is the air temperature when a cricket makes 120 chirps per minute?

4. a. According to the formula, how many chirps would a cricket make at 35°F?

   b. Does this number make sense? Explain.

Challenge: Skills and Applications

For use with pages 144–149

In Exercises 1-10, solve the equation.

1. $\frac{2}{3}x - \frac{3}{5} = \frac{4}{9}$

2. $1 - \frac{8}{9}x = \frac{5}{6}$

3. $8 - 2|r| = -12$

4. $3|y| - 22 = 0$

5. $\frac{|x| - 5}{4} = 6$

6. $6 - \frac{2}{3}|n| = -8$

7. $33 - \frac{1}{5}(3x + 1) = 34$

8. $\frac{1}{2} - 3(x + 1) = 8$

9. $6a - 4[2 - 3(4a - 3)] = -17$

10. $5|g| - (4 - 3|g|) = 20$

11. Marianna Montaine has 15 grams of plant nutrient that she is testing in a laboratory. She needs to give each plant in the test 0.4 grams of nutrient, and she wants to have 3 grams left over for future use. Write an algebraic model for this situation, where x represents the number of plants Marianna can test. Solve the problem.

12. A hexagon has three sides of the same length, two sides which are each two-thirds that length, and a sixth side of length $14\frac{1}{2}$ centimeters. The perimeter of the hexagon is 86 centimeters. Write an algebraic model for this situation, where x represents the length of the three equal sides. Solve the equation. Then find the length of each side.

13. A person has quarters, dimes, nickels, and pennies, with a total value of $3.86. The number of nickels is twice the number of quarters. The number of quarters is two less than the number of dimes. There are 40 coins altogether. Write and solve an equation to find the number of each coin.

NAME _____ DATE _____

Quiz 1

For use after Lessons 3.1–3.3

Solve the equation. *(Lesson 3.1–3.3)*

1. $15 = t + 23$

2. $a - (-4) = -35$

3. $12m = 4$

4. $-7 = -\dfrac{1}{6}x$

5. $\dfrac{3}{4}n + 16 = 19$

6. $-3(5 - y) = -60$

7. $\dfrac{2}{3}(x - 8) = 4$

Answers
1. _____
2. _____
3. _____
4. _____
5. _____
6. _____
7. _____

TEACHER'S NAME _____ CLASS _____ ROOM _____ DATE _____

Lesson Plan

2-day lesson (See *Pacing the Chapter,* TE page 128A)   **For use with pages 150–156**

GOAL **Solve equations that have variables on both sides.**

State/Local Objectives _____

_____

✓ **Check the items you wish to use for this lesson.**

STARTING OPTIONS
____ Homework Check (3.3): TE page 147; Answer Transparencies
____ Homework Quiz (3.3): TE page 149, CRB page 48, or Transparencies
____ Warm-Up: CRB page 48 or Transparencies

TEACHING OPTIONS
____ Developing Concepts: SE page 150; CRB page 49 (Activity Support Master)
____ Lesson Opener: CRB page 50 or Transparencies
____ Investigation 4: Algebra Tiles Investigations
____ Examples Day 1: 1–3, SE pages 151–152; Day 2: 4, SE page 153
____ Extra Examples: TE pages 152–153 or Transparencies; Internet Help at *www.mcdougallittell.com*
____ Checkpoint Exercises: Day 1: Exs. 1–6, SE page 152; Day 2: Exs. 6–9, SE page 153
____ Concept Check: TE page 153
____ Guided Practice Exercises: SE page 154 Day 1: Exs. 1–8; Day 2: Exs. 9–16

APPLY/HOMEWORK
Homework Assignment
____ Transitional: Day 1: EP p. 90 Exs. 48–53; pp. 154–156, Exs. 17–24, 35, 47, 57–85 odd;
      Day 2: pp. 155–156 Exs. 37, 38, 43–44, 54–56, 58–86 even
____ Average: Day 1: pp. 154–156 Exs. 25–30, 36, 48, 49, 58–60, 64–66;
      Day 2: pp. 155–156 Exs. 39, 40, 45, 46, 54–57, 71–73, 81–83
____ Advanced: Day 1: pp. 154–156, Exs. 29–36, 47–49, 61–63, 74–77;
      Day 2: pp. 155–156 Exs. 41, 42, 45, 46, 50–56*, 84–86; EC: CRB p. 58

Reteaching the Lesson
____ Practice Masters: CRB pages 51–52 (Level A, Level B)
____ Reteaching with Practice: CRB pages 53–54 or Practice Workbook with Examples;
      Resources in Spanish
____ Personal Student Tutor: CD-ROM

Extending the Lesson
____ Learning Activity: CRB page 56
____ Interdisciplinary/Real-Life Applications: CRB page 57
____ Challenge: CRB page 58

ASSESSMENT OPTIONS
____ Daily Quiz (3.4): TE page 156, CRB page 61, or Transparencies
____ Standardized Test Practice: SE page 156; STP Workbook; Transparencies

Notes _____

_____

_____

Lesson Plan for Block Scheduling

1-block lesson (See *Pacing the Chapter*, TE page 128A)   **For use with pages 150–156**

GOAL   **Solve equations that have variables on both sides.**

State/Local Objectives _____

_____

✓ **Check the items you wish to use for this lesson.**

STARTING OPTIONS

____ Homework Check (3.3): TE page 147; Answer Transparencies
____ Homework Quiz (3.3): TE page 149, CRB page 48, or
      Transparencies
____ Warm-Up: CRB page 48 or Transparencies

TEACHING OPTIONS

____ Developing Concepts: SE page 150; CRB page 49 (Activity Support Master)
____ Lesson Opener: CRB page 50 or Transparencies
____ Investigation 4: Algebra Tiles Investigations
____ Examples: 1–4, SE pages 151–153
____ Extra Examples: TE pages 152–153 or Transparencies; Internet Help at *www.mcdougallittell.com*
____ Checkpoint Exercises: Exs. 1–6, SE page 152; Exs. 6–9, SE page 153
____ Concept Check: TE page 153
____ Guided Practice Exercises: SE page 154 Day 1: Exs. 1–8; Day 2: Exs. 9–16

APPLY/HOMEWORK

Homework Assignment

____ Block Schedule: pp. 154–156 Exs. 25–30, 36, 39, 40, 45, 46, 48, 49, 54–60, 64–66, 71–73, 81–83

Reteaching the Lesson

____ Practice Masters: CRB pages 51–52 (Level A, Level B)
____ Reteaching with Practice: CRB pages 53–54 or Practice Workbook with Examples;
      Resources in Spanish
____ Personal Student Tutor: CD-ROM

Extending the Lesson

____ Learning Activity: CRB page 56
____ Interdisciplinary/Real-Life Applications: CRB page 57
____ Challenge: CRB page 58

ASSESSMENT OPTIONS

____ Daily Quiz (3.4): TE page 156, CRB page 61, or Transparencies
____ Standardized Test Practice: SE page 156; STP Workbook; Transparencies

Notes _____

_____

CHAPTER PACING GUIDE	
Day	**Lesson**
1	3.1 (all)
2	3.2 (all); 3.3 (all)
3	**3.4 (all)**
4	3.5 (all); 3.6 (begin)
5	3.6 (end); 3.7 (all)
6	3.8 (all); 3.9 (all)
7	Ch. 3 Review and Assess

Lesson 3.4

Algebra 1
Chapter 3 Resource Book

WARM-UP EXERCISES

For use before Lesson 3.4, pages 150–156

Simplify.

1. $2(4x - 5)$

2. $6x - 3x + 5$

3. $-7(2 - 5x)$

4. $3(x - 2) + 5x$

5. $9x - (4 - 3x)$

....

DAILY HOMEWORK QUIZ

For use after Lesson 3.3, pages 144–149

Solve the equation.

1. $4x + 5 = 13$

2. $8t - 3t = 30$

3. $\frac{3}{4}(k - 3) = 9$

4. $a - (6a + 5) = 45$

5. A student works at a job which pays $6.00 per hour.
This week the student also received a bonus of $50. If the
total pay for the week was $284, how many hours did the
student work? Use the verbal model to write an equation
to solve the problem.

$$\boxed{\begin{array}{c}\text{Hourly}\\\text{rate}\end{array}} \times \boxed{\begin{array}{c}\text{Hours}\\\text{worked}\end{array}} + \boxed{\text{Bonus}} = \boxed{\begin{array}{c}\text{Total}\\\text{pay}\end{array}}$$

Algebra 1
Chapter 3 Resource Book

Lesson 3.4

NAME _____ DATE _____

Activity Support Master

For use with page 150

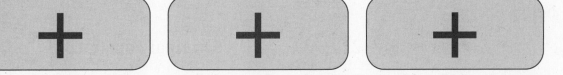

Algebra 1
Chapter 3 Resource Book

49

NAME _____ DATE _____

Activity Lesson Opener

For use with pages 150–156

SET UP: Work with a partner.

1. In the two equations below, circle all terms that contain variables. How do these equations differ from those you have already learned to solve?

$$4x + 5 = 2x - 3 \qquad\qquad 10m - 16 = 2m$$

2. Complete the steps to solve the equation $4x + 5 = 2x - 3$. Explain what happens in each step. Use the questions in the first two steps to help you.

$$4x + 5 = 2x - 3 \qquad \text{Original equation}$$

a. $\quad 4x + 5 - 2x = 2x - 3 - 2x \qquad$ What was done to both sides?

b. $\quad (4x - 2x) + 5 = (2x - 2x) - 3 \qquad$ What are the terms inside the parentheses called?

c. $\quad \underline{} + 5 = -3$

d. $2x + 5 - \underline{} = -3 - \underline{}$

e. $\quad\qquad 2x = \underline{}$

f. $\quad\qquad \dfrac{2x}{} = \dfrac{-8}{}$

g. $\quad\qquad x = \underline{}$

3. Complete the steps to solve the equation $10m - 16 = 2m$. Explain what happens in each step.

a. $\quad 10m - 16 - 10m = 2m - 10m \qquad$ What was done to both sides?

b. $(10m - 10m) - 16 = (2m - 10m)$

c. $\quad\qquad -16 = \underline{}$

d. $\quad\qquad \dfrac{-16}{} = \dfrac{-8m}{}$

e. $\quad\qquad \underline{} = m$

NAME _____ DATE _____

Practice A

For use with pages 150-156

Describe each step used in solving the equation.

1. $9x - 4 = 7x + 8$ **2.** $-4x + 9 = 2x + 3$ **3.** $4(2x - 9) = 4$

a. $2x - 4 = 8$ **a.** $-4x + 6 = 2x$ **a.** $8x - 36 = 4$

b. $2x = 12$ **b.** $6 = 6x$ **b.** $8x = 40$

c. $x = 6$ **c.** $1 = x$ **c.** $x = 5$

Solve the equation and describe each step you use.

4. $2x = x + 9$ **5.** $4x - 6 = 3x$ **6.** $-2x = -3x + 8$

7. $7x = 5x + 24$ **8.** $7x + 5 = 6x$ **9.** $12x = 9x - 15$

Solve the equation if possible.

10. $2x + 5 = 3x$ **11.** $-2x = -4x + 20$ **12.** $7x - 20 = -3x$

13. $7x = 4x - 9$ **14.** $-8x - 70 = 2x$ **15.** $8x - 3 = 8x$

16. $3(x - 1) = 3x - 3$ **17.** $2x + 3 = 4x + 5$ **18.** $-3x - 4 = 4x + 10$

19. $8x - 3 = 19 + 5x$ **20.** $x = 21 - 2x$ **21.** $x + 12 = -x$

In Exercises 22–24, write and solve an equation to answer the question.

22. *Dimensions of a Circular Flower Garden*
A flower garden has the shape pictured
below. The diameter of the outer circle is
twice the diameter of the inner circle. The
lengths of the walkways are each 6 feet long.
What is the diameter of the inner circle?

23. *Balanced Scale*   On one side of a scale
there are 6 blocks, 3 weighing 2 grams each
and 3 weighing x grams each. The scale is
balanced if 5 blocks weighing x grams each
are placed on the other side of the scale.
How much does each of the unknown blocks
weigh?

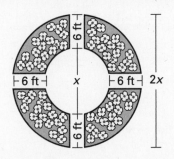

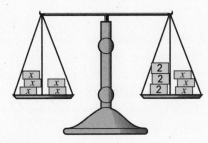

24. *Distance-Rate-Time*   Two cars travel the same distance. The first
car travels at a rate of 40 miles per hour and reaches its destination in
t hours. The second car travels at a rate of 55 miles per hour and reaches
its destination 3 hours earlier than the first car. How long does it take
for the first car to reach its destination?

$$\boxed{\text{Rate of car 1}} \cdot \boxed{\text{Time for car 1}} = \boxed{\text{Rate of car 2}} \cdot \boxed{\text{Time for car 2}}$$

NAME _____ DATE _____

Practice B

For use with pages 150–156

Solve the equation and describe each step you use.

1. $2x = -x + 9$
2. $-4x - 6 = 3x + 1$
3. $5 - 2x = 3x + 8$
4. $40 - 4x = -7x + 7$
5. $6x - 2 = -4x + 8x$
6. $-(18 + x) = 22 - 12x + x$

Solve the equation if possible.

7. $5x + 5 = 6x$
8. $-2x = -4x + 24$
9. $7x - 40 = -3x$
10. $7x = 4x - 15$
11. $-8x - 70 = 6x$
12. $8x - 9 = 8x$
13. $2(2x - 3) = 4x - 6$
14. $-3 - (-4x) = -4x + 5$
15. $-(10 - x) = 3x + 12$
16. $8x - 4 = 19 + 5x$
17. $x = 35 - 4x$
18. $x + 48 = -x$
19. $\frac{1}{2}x - 8 = 14 + \frac{1}{2}x$
20. $2(2x - 6) = 8x$
21. $2(6x - 9) = 12x - 18$

In Exercises 22–24, write and solve an equation to answer the question.

22. *Dimensions of a Circular Flower Garden*
A flower garden has the shape pictured
below. The diameter of the inner circle is 12
feet. What are the lengths of the walkways?

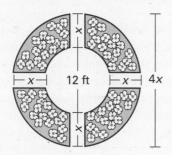

23. *Balanced Scale*   On one side of a scale
there are 4 blocks, 2 weighing 2 grams each
and 2 weighing x grams each. The scale is
balanced if 8 blocks weighing x grams each
are placed on the other side of the scale.
How much does each of the unknown blocks
weigh?

24. *Distance-Rate-Time*   Two cars travel the same distance. The first
car travels at a rate of 35 miles per hour and reaches its destination in
t hours. The second car travels at a rate of 50 miles per hour and reaches
its destination 3 hours earlier than the first car. How long does it take for
the first car to reach its destination? How long does it take for the second
car to reach its destination?

Rate of car 1	·	Time for car 1	=	Rate of car 2	·	Time for car 2

25. *Extension*   Write an equation that has no solution.

26. *Extension*   Write an equation that is an identity.

NAME _____ DATE _____

Reteaching with Practice

For use with pages 150–156

GOAL Solve equations that have variables on both sides.

> **VOCABULARY**
>
> An **identity** is an equation that is true for all values of the variable.

EXAMPLE 1 *Collecting Variables on One Side*

Solve $20 - 3x = 2x$.

SOLUTION

Think of $20 - 3x$ as $20 + (-3x)$. Since $2x$ is greater than $-3x$, collect the x-terms on the right side.

$20 - 3x = 2x$	Write original equation.
$20 - 3x + 3x = 2x + 3x$	Add $3x$ to each side.
$20 = 5x$	Simplify.
$\dfrac{20}{5} = \dfrac{5x}{5}$	Divide each side by 5.
$4 = x$	Simplify.

Exercises for Example 1

Solve the equation.

1. $5q = -7q + 6$ **2.** $14d - 6 = 17d$ **3.** $-y + 7 = -8y$

EXAMPLE 2 *Identifying the Number of Solutions*

a. Solve $2x + 3 = 2x + 4$.   **b.** Solve $-(t + 5) = -t - 5$.

SOLUTION

a.	$2x + 3 = 2x + 4$	Write original equation.
	$2x + 3 - 3 = 2x + 4 - 3$	Subtract 3 from each side.
	$2x = 2x + 1$	Simplify.
	$0 = 1$	Subtract $2x$ from each side.

The original equation has *no solution*, because $0 \neq 1$ for any value of x.

b.	$-(t + 5) = -t - 5$	Write original equation.
	$-t - 5 = -t - 5$	Use distributive property.
	$-5 = -5$	Add t to each side.

All values of t are solutions, because $-5 = -5$ is always true.
The original equation is an *identity*.

LESSON

3.4

CONTINUED

NAME _____ DATE _____

Reteaching with Practice

For use with pages 150–156

Exercises for Example 2

Identify the number of solutions.

4. $9z - 3 = 9z$ **5.** $2(f - 7) = 2f - 14$ **6.** $n + 3 = -5n$

EXAMPLE 3 *Solving Real-Life Problems*

A health club charges nonmembers $2 per day to swim and $5 per day for aerobics classes. Members pay a yearly fee of $200 plus $3 per day for aerobics classes. Write and solve an equation to find the number of days you must use the club to justify a yearly membership.

SOLUTION

Let n represent the number of days that you use the club. Then find the number of times for which the two plans would cost the same.

$2n + 5n = 200 + 3n$	Write equation.
$7n = 200 + 3n$	Combine like terms.
$7n - 3n = 200 + 3n - 3n$	Subtract $3n$ from each side.
$4n = 200$	Simplify.
$\dfrac{4n}{4} = \dfrac{200}{4}$	Divide each side by 4.
$n = 50$	Simplify.

You must use the club 50 days to justify a yearly membership.

Exercises for Example 3

7. Rework Example 3 if nonmembers pay $3 per day to swim.

8. Rework Example 3 if members pay a yearly fee of $220.

Lesson 3.4

NAME _____ DATE _____

Quick Catch-Up for Absent Students

For use with pages 150–156

The items checked below were covered in class on (date missed) _____

Developing Concepts Activity: Variables on Both Sides (p. 150)

____ **Goal:** Use algebra tiles to solve equations with variables on both sides.

Lesson 3.4: Solving Equations with Variables on Both Sides

____ **Goal:** Solve equations that have variables on both sides.

Material Covered:

____ Student Help: Study Tip

____ Example 1: Collect Variables on Left Side

____ Student Help: Look Back

____ Example 2: Collect Variables on Right Side

____ Example 3: Combine Like Terms First

____ Example 4: Identify Number of Solutions

Vocabulary:

identity, p. 153

____ Other (specify)_____

_____

Homework and Additional Learning Support

____ Textbook (specify) _pp. 154–156_____

_____

____ Internet: Extra examples at www.mcdougallittell.com

____ *Reteaching with Practice* worksheet (specify exercises) _____

____ *Personal Student Tutor* for Lesson 3.4

NAME _____ DATE _____

Learning Activity

For use with pages 150–156

GOAL **Explain how to solve linear equations to a friend.**

Materials: Textbook

Solving Linear Equations

As you have learned to solve linear equations, it is important that you be able to explain your solutions to someone else. In this activity, you will explain how to solve a linear equation to a friend who has missed a few days of school. You will then work with a partner to evaluate one another's work.

Instructions

1 Assume that a friend of yours has been sick from school and missed the lessons on linear equations. This friend calls you to ask for his or her math assignment.

2 Your friend is confused by the lessons that he or she has missed and asks you to explain the procedure used to solve linear equations.

3 Choose one of the exercises from the Practice and Application section on page 154 of your text. Write the dialogue you would use to explain how to solve that exercise to your friend.

4 Exchange papers with your partner.

Analyzing the Results

1. Were you able to explain the solution accurately? Would someone really understand what you wrote?

2. Were you able to understand your partner's explanation? Was it accurate?

3. In your opinion, what is the hardest thing about solving linear equations? Make sure that you speak to your teacher if you are having problems.

NAME _____ DATE _____

Real-Life Application:
When Will I Ever Use This?

For use with pages 150–156

Recycling

In 1998, with a 62.8% recycling rate, the aluminum can was the most prevalent recycled packaging container in the United States. The benefits to recycling aluminum cans are numerous. The environment benefits from conservation of valuable energy and natural resources, as well as reduced landfill space. The volume of aluminum cans reaching a landfill today makes up only 1.1% of the total consumer waste. The aluminum industry itself also benefits, for recycling requires 95% less energy than producing aluminum from the ore and creates a large portion of the actual metal supply. Finally, the consumer benefits, with earnings of $990 million in 1998 alone. The money paid for the aluminum provides personal income for individuals, funds for charities, and often operating revenues for city-sponsored recycling programs.

The aluminum industry has also helped the recycling effort by continuously decreasing the weight and increasing the recycled content of aluminum cans. In 1998, the number of cans per pound was up to approximately 33 and the recycled content of a can was about 51%. Although the industry's ultimate goal is to recycle 100% of the aluminum cans, the progress made in the last 25 years is amazing!

In Exercises 1–4, use the following information.

You decide to recycle aluminum cans to make some extra spending money. The Cans-for-Cash Recycling Center pays $0.48 per pound of aluminum cans. The Earth Saver Recycling Center pays $0.42 per pound of aluminum cans and a flat fee of $1.80 if you crush the cans.

1. Write an equation to help you decide which recycling center you will go to.

2. Solve the equation to find the weight at which you would receive the same amount at either recycling center.

3. You have 10 pounds of recycled aluminum cans. Which recycling center would you go to? Explain.

4. You have 40 pounds of recycled aluminum cans. Which recycling center would you go to? Explain.

Challenge: Skills and Applications

For use with pages 150–156

For Exercises 1–6, solve the equation.

1. $7 + \frac{2}{3}y = \frac{1}{6}y - 5$

2. $2 + \frac{4}{3}n = 11 - \frac{1}{6}n$

3. $\frac{1}{2}(r + 1) = \frac{2}{7}(r + 14)$

4. $-\frac{2}{3}h + 14 = h - 6$

5. $\frac{1}{2}(4x - 5) - 1\frac{1}{2} = \frac{2}{3}(3x - 6)$

6. $\frac{3}{4}\left(\frac{8}{3}x - 8\right) - 3 = \frac{1}{2}(4x + 6)$

For Exercises 7–9, solve the equation.

Example: $\frac{5}{2x} + 3 = \frac{7}{x}$

Solution: $x\left(\frac{5}{2x} + 3\right) = x\left(\frac{7}{x}\right)$ Multiply each side by x.

$\frac{5}{2} + 3x = 7$ Distribute the x.

$\frac{5}{2} + 3x - \frac{5}{2} = 7 - \frac{5}{2}$ Subtract $\frac{5}{2}$ from each side.

$3x = \frac{9}{2}$ Simplify.

$x = \frac{3}{2}$ Divide each side by 3.

7. $\frac{2}{x} - 4 = \frac{3}{x}$ 8. $6 - \frac{5}{x} = \frac{1}{2x}$ 9. $\frac{4}{3x} + 5 = -\frac{2}{x}$

For Exercises 10–13, use the following information.

His new employer has offered Malcolm Davis a choice of profit-sharing plans. He can receive $\frac{1}{90}$ of the company's gross income or he can receive $\frac{1}{60}$ of the company's profit. Gross income is the total amount the company takes in. Profit is the income that is left after the expenses have been subtracted off. The company's expenses for one month are $100,000.

10. Use a verbal model to write an equation for finding the gross income that would give Malcolm Davis the same amount of money with either plan.

11. Solve the equation from Exercise 10 and interpret the solution.

12. How much would Malcolm Davis receive when the two plans are the same?

13. If the gross income of the company is less than the amount from Exercise 11, which plan would be better for Malcolm Davis?

TEACHER'S NAME _____ CLASS _____ ROOM _____ DATE _____

Lesson Plan

1-day lesson (See *Pacing the Chapter*, TE page 128A) For use with pages 157–162

GOAL **Solve more complicated equations that have variables on both sides.**

State/Local Objectives _____

_____

✓ Check the items you wish to use for this lesson.

STARTING OPTIONS
____ Homework Check (3.4): TE page 154; Answer Transparencies
____ Homework Quiz (3.4): TE page 156, CRB page 61, or Transparencies
____ Warm-Up: CRB page 61 or Transparencies

TEACHING OPTIONS
____ Lesson Opener: CRB page 62 or Transparencies
____ Examples 1–4: SE pages 157–159
____ Extra Examples: TE pages 158–159 or Transparencies; Internet Help at *www.mcdougallittell.com*
____ Checkpoint Exercises: SE pages 158–159
____ Concept Check: TE page 159
____ Guided Practice Exercises: SE page 160

APPLY/HOMEWORK
Homework Assignment
____ Transitional: pp. 160–162 Exs. 19–24, 31, 32, 37, 38, 40, 43, 52, 53, 55–69 odd
____ Average: pp. 160–162 Exs. 25–28, 33, 34, 39–42, 52, 58–60, 64–66
____ Advanced: pp. 160–162 Exs. 29, 30, 35, 36, 43–53*, 61–63, 67–69; EC: CRB p. 69

Reteaching the Lesson
____ Practice Masters: CRB pages 63–64 (Level A, Level B)
____ Reteaching with Practice: CRB pages 65–66 or Practice Workbook with Examples;
     Resources in Spanish
____ Personal Student Tutor: CD-ROM

Extending the Lesson
____ Interdisciplinary/Real-Life Applications: CRB page 68
____ Challenge: CRB page 69

ASSESSMENT OPTIONS
____ Daily Quiz (3.5): TE page 162, CRB page 72, or Transparencies
____ Standardized Test Practice: SE page 162; STP Workbook; Transparencies

Notes _____

_____

_____

TEACHER'S NAME _____ CLASS _____ ROOM _____ DATE _____

Lesson Plan for Block Scheduling

Half-block lesson (See *Pacing the Chapter,* TE page 128A)   For use with pages 157–162

GOAL   Solve more complicated equations that have variables on both sides.

State/Local Objectives _____

_____

_____

✓ **Check the items you wish to use for this lesson.**

STARTING OPTIONS

____ Homework Check (3.4): TE page 154; Answer Transparencies

____ Homework Quiz (3.4): TE page 156,

CRB page 61, or Transparencies

____ Warm-Up: CRB page 61 or Transparencies

TEACHING OPTIONS

____ Lesson Opener: CRB page 62 or Transparencies

____ Examples 1–4: SE pages 157–159

____ Extra Examples: TE pages 158–159 or Transparencies; Internet Help at *www.mcdougallittell.com*

____ Checkpoint Exercises: SE pages 158–159

____ Concept Check: TE page 159

____ Guided Practice Exercises: SE page 160

APPLY/HOMEWORK

Homework Assignment (See also the assignment for Lesson 3.6.)

____ Block Schedule: pp. 160–162: Exs. 25–28, 33, 34, 39–42, 52, 58–60, 64–66

Reteaching the Lesson

____ Practice Masters: CRB pages 63–64 (Level A, Level B)

____ Reteaching with Practice: CRB pages 65–66 or Practice Workbook with Examples; Resources in Spanish

____ Personal Student Tutor: CD-ROM

Extending the Lesson

____ Interdisciplinary/Real-Life Applications: CRB page 68

____ Challenge: CRB page 69

ASSESSMENT OPTIONS

____ Daily Quiz (3.5): TE page 162, CRB page 72, or Transparencies

____ Standardized Test Practice: SE page 162; STP Workbook; Transparencies

Notes _____

_____

_____

_____

CHAPTER PACING GUIDE	
Day	**Lesson**
1	3.1 (all)
2	3.2 (all); 3.3 (all)
3	3.4 (all)
4	**3.5 (all);** 3.6 (begin)
5	3.6 (end); 3.7 (all)
6	3.8 (all); 3.9 (all)
7	Ch. 3 Review and Assess

LESSON

3.5

NAME _____ DATE _____

Available as
a transparency

Lesson 3.5

WARM-UP EXERCISES

For use before Lesson 3.5, pages 157–162

Solve.

1. $8x + 3 - 5x = 18$

2. $4(x + 2) = -16$

3. $7(3 - 2x) + 9x = 6$

4. $25x + 10 = 30x + 7$

· ·

DAILY HOMEWORK QUIZ

For use after Lesson 3.4, pages 150–156

Solve the equation.

1. $18 + 3g = 5g$

2. $8x - 2x = 3x + 21$

3. $t + 3 - 2t = 4t + 1$

4. $-r + 7 - 8r = 11 - 10r$

5. A car is traveling on a highway at 50 miles per hour at the
time that a second car enters the highway 2 miles behind
the first car. The second car is traveling 60 miles per hour
and is traveling in the same direction as the first car. How
long will it take the second car to catch up to the first car?
Solve the equation $50t = 60t - 2$, where t is the time in
hours it takes the second car to catch up to the first.

Visual Approach Lesson Opener

For use with pages 157–162

1. The graph shows the total amount of money you have earned so far this year. For example, the point $(4, 200)$ shows that in the first 4 weeks you earned a total of $200.

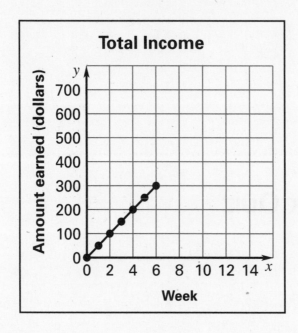

a. What does the graph tell you about how much you earn each week?

b. You want to find how many weeks it will take you to earn $750. Write an equation to model this situation.

2. The diagram shows the layout of a rectangular dog run that needs to be fenced in.

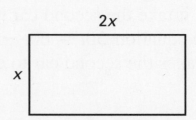

a. Describe the dimensions of the dog run in words.

b. You have 30 feet of fencing and you plan to use it all. Write an equation that models this situation.

NAME _____ DATE _____

Practice A

For use with pages 157–162

Solve the equation.

1. $4(x + 7) = -9(x - 6)$

2. $-(2y + 1) = 8(y + 2) - 7$

3. $-5(2x + 6) = -3(-4 + x)$

4. $3a + 2(a - 1) = 7a - 4(a + 2)$

5. $9(k - 4) + 2k = 2(k - 1) + 7k$

6. $3c - 7(-c + 4) = -2(c - 5) + 6c$

7. $4(a + 8) = \frac{1}{2}(4a - 8)$

8. $\frac{1}{3}(18x + 15) = \frac{1}{4}(4x + 12)$

9. $-\frac{4}{5}(15y - 20) = 2(2y + 8)$

10. $9(20 - 12t) = 8(6 + 9t) - 24t$

Yearbook Layout **In Exercises 11–13, use the following information.**

A page of a school yearbook is $8\frac{1}{2}$ inches by 11 inches. The left and right margins are 1 inch and $2\frac{1}{2}$ inches, respectively. The space between pictures is $\frac{1}{4}$ inch. How wide can each picture be to fit 3 across the width of the page?

11. Write a verbal model for this problem.

12. Write an equation for the model.

13. Solve the equation and answer the question.

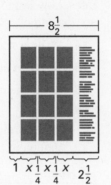

Saving and Spending **In Exercises 14–17, use the following information.**

Currently, you have $60 and your sister has $135. You decide to save $5 of your allowance each week, while your sister decides to spend her whole allowance plus $10 each week. How long will it be before you have as much money as your sister?

14. Write a verbal model for this problem.

15. Write an equation for the model.

16. Solve the equation and answer the question.

17. Copy and complete the table below using the information from the original problem statement.

Week	0	1	2	3	4	5
Your money						
Sister's money						

NAME _____ DATE _____

Practice B

For use with pages 157–162

Solve the equation.

1. $-2(x + 10) = 3(2x - 4)$

2. $-(2y + 1) = 8(y + 2) - 7$

3. $-4(-5x + 8) = 7(-6 + 2x)$

4. $7t + 6(t - 2) = 4t + 10(2t + 1)$

5. $12(y - 1) - 4y = 3(2y - 2) + 9y$

6. $3c - 7(-c + 4) = -2(c - 5) + 6c$

7. $-8(x + 3) = \frac{1}{2}(-6x - 18)$

8. $\frac{3}{5}(25x + 10) = 6x + 15$

9. $-8(45y + 18) = 3(12y + 18)$

10. $3(20 - 12t) = \frac{8}{7}(14 + 21t) - 12t$

Cassette Storage **In Exercises 11–13, use the following information.**

You have a box that is a good size for your tape collection. Two rows of tapes will fit in the box. The box is 10 inches wide. Each tape is $\frac{5}{8}$ inches wide. How many tapes will fit in the box?

11. Write a verbal model for this problem.

12. Write an equation for the model.

13. Solve the equation and answer the question.

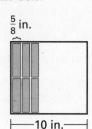

$\frac{5}{8}$ in.

⊢——10 in.——⊣

Temperature Change **In Exercises 14–18, use the following information.**

In Detroit the temperature is 69° F and is rising at a rate of 2° F per hour. In Atlanta the temperature is 84° F and is falling at a rate of 3° F per hour. If the temperatures continue to change at the same rates, how long will it be before the temperatures are the same?

14. Write a verbal model for this problem.

15. Write an equation for the model.

16. Solve the equation and answer the question.

17. Copy and complete the table below using the information from the original problem statement.

Hour	0	1	2	3
Detroit temperature (°F)				
Atlanta temperature (°F)				

18. Use the graph to check the answer. Is the solution correct? Explain.

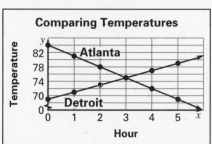

Reteaching with Practice

For use with pages 157–162

GOAL Solve more complicated equations that have variables on both sides.

EXAMPLE 1 *Solving a More Complicated Equation*

Solve the equation.

a. $2(x - 5) + 10 = -(-3x + 2)$

b. $-3(5x + 1) + 2x = 2(4x - 3)$

c. $\frac{1}{2}(16 - 6x) = 15 - \frac{1}{3}(9 + 15x)$

SOLUTION

a.

$2(x - 5) + 10 = -(-3x + 2)$	Write original equation.
$2x - 10 + 10 = 3x - 2$	Use distributive property.
$2x = 3x - 2$	Combine like terms.
$-x = -2$	Subtract $3x$ from each side.
$x = 2$	Divide each side by -1.

The solution is 2. Check this in the original equation.

b.

$-3(5x + 1) + 2x = 2(4x - 3)$	Write original equation.
$-15x - 3 + 2x = 8x - 6$	Use distributive property.
$-13x - 3 = 8x - 6$	Combine like terms.
$-21x - 3 = -6$	Subtract $8x$ from each side.
$-21x = -3$	Add 3 to each side.
$x = \frac{1}{7}$	Divide each side by -21.

The solution is $\frac{1}{7}$. Check this in the original equation.

c.

$\frac{1}{2}(16 - 6x) = 15 - \frac{1}{3}(9 + 15x)$	Write original equation.
$8 - 3x = 15 - 3 - 5x$	Use distributive property.
$8 - 3x = 12 - 5x$	Combine like terms.
$8 + 2x = 12$	Add $5x$ to each side.
$2x = 4$	Subtract 8 from each side.
$x = 2$	Divide each side by 2.

The solution is 2. Check this in the original equation.

Exercises for Example 1

1. $2(x + 5) + 3x = 3(-2x - 1)$

2. $\frac{3}{4}(8x - 20) = 6x + 12 - 12x$

NAME _____ DATE _____

Reteaching with Practice

For use with pages 157–162

EXAMPLE 2 *Drawing a Diagram*

The front page of your school newspaper is $11\frac{1}{4}$ inches wide. The left margin is 1 inch and the right margin is $1\frac{1}{2}$ inches. The space between the four columns is $\frac{1}{4}$ inch. Find the width of each column.

SOLUTION

The diagram shows that the page is made up of the width of the left margin, the width of the right margin, three spaces between the columns, and the four columns.

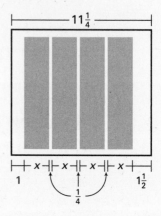

Verbal Model	Left margin	+	Right margin	+ 3 ·	Space between columns	+ 4 ·	Column width	=	Page width

Labels   Left margin = 1   (inch)

Right margin = $1\frac{1}{2}$ (inches)

Space between columns = $\frac{1}{4}$ (inch)

Column width = x (inches)

Page width = $11\frac{1}{4}$ (inches)

Algebraic Model $1 + 1\frac{1}{2} + 3\left(\frac{1}{4}\right) + 4x = 11\frac{1}{4}$

Solving for x, you find that each column can be 2 inches wide.

Exercise for Example 2

3. Rework Example 1 if the front page of the newspaper has three columns.

Lesson 3.5

NAME _____ DATE _____

Quick Catch-Up for Absent Students

For use with pages 157–162

The items checked below were covered in class on (date missed) _____

Lesson 3.5: More on Linear Equations

_____ **Goal:** Solve more complicated equations that have variables on both sides.

Material Covered:

_____ Student Help: Study Tip

_____ Example 1: Solve a More Complicated Equation

_____ Student Help: Study Tip

_____ Example 2: Solve a More Complicated Equation

_____ Example 3: Solve a More Complicated Equation

_____ Example 4: Compare Payment Plans

_____ Other (specify) _____

_____

Homework and Additional Learning Support

_____ Textbook (specify) pp. 160–162 _____

_____

_____ Internet: Extra examples at www.mcdougallittell.com

_____ *Reteaching with Practice* worksheet (specify exercises)_____

_____ *Personal Student Tutor* for Lesson 3.5

LESSON 3.5

Real-Life Application: When Will I Ever Use This?

For use with pages 157–162

Tunnels

Tunnels have been used throughout history for various purposes, such as irrigation and transportation.

Boring Brothers, Inc. has a contract to dig a tunnel through a mountain to accommodate the construction of a major highway. Crew A starts at the west end and digs at a rate of 9 meters per day. Crew B starts at the east end two days after Crew A and digs at a rate of 12 meters per day.

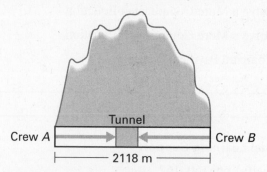

In Exercises 1–4, use the information above.

1. Let x be the number of days Crew A has been digging. Write an expression for the number of meters Crew A has dug after x days.

2. In terms of x, how many days has Crew B been digging? Write an expression for the number of meters Crew B has dug in this number of days.

3. Write and solve an equation to find how many days it will take for both crews to dig the same number of meters.

4. The total length of the tunnel is to be 2118 meters. Write and solve an equation to find how many days it takes to dig the tunnel from the time Crew A starts.

Algebra 1
Chapter 3 Resource Book

NAME _____ DATE _____

Challenge: Skills and Applications

For use with pages 157–162

For Exercises 1–5, use the following information. Check your answer with a table.

Two grocery stores sell rice in bulk. The first charges $0.55 per pound. The second charges $0.75 per pound for up to 3 pounds and $0.40 per pound for anything over 3 pounds.

1. Write expressions for the cost of rice at each store in terms of the number of pounds bought, assuming you buy more than 3 pounds.

2. Write and solve an equation that relates the two expressions from Exercise 1.

3. Interpret your result from Exercise 2.

4. Evaluate each expression from Exercise 1 for the value of x from Exercise 2. Interpret the result.

5. Describe what happens for values of x less than the one found in Exercise 2 and for values greater than it.

In Exercises 6–9, you will create your own word problem and then represent the situation three ways.

6. Write a word problem for which drawing a diagram would help the reader to understand the relationships.

7. Draw a diagram to visualize the problem you wrote in Exercise 6.

8. Use your diagram from Exercise 7 to write a verbal model for the situation.

9. Use the verbal model from Exercise 8 to write an algebraic model.

Lesson Plan

2-day lesson (See *Pacing the Chapter*, TE page 128A)   For use with pages 163–170

GOAL **Find exact and approximate solutions of equations that contain decimals.**

State/Local Objectives _____

_____

✓ **Check the items you wish to use for this lesson.**

STARTING OPTIONS

_____ Homework Check (3.5): TE page 160; Answer Transparencies

_____ Homework Quiz (3.5): TE page 162, CRB page 72, or Transparencies

_____ Warm-Up: CRB page 72 or Transparencies

TEACHING OPTIONS

_____ Lesson Opener: CRB page 73 or Transparencies

_____ Examples Day 1: 1–2, SE pages 163–164; Day 2: 3–4, SE pages 164–165

_____ Extra Examples: TE pages 164–165 or Transparencies; Internet Help at *www.mcdougallittell.com*

_____ Checkpoint Exercises: Day 1: Exs. 1–7, SE pages 163–164; Day 2: Ex. 8, SE page 165

_____ Technology Graphing Calculator: Solving Multi-Step Equations: SE page 170; CRB page 75 (Keystrokes)

_____ Graphing Calculator Activity with Keystrokes: CRB pages 74–75

_____ Concept Check: TE page 165

_____ Guided Practice Exercises: SE page 166; Day 1: Exs. 1–18; Day 2: Ex. 19

APPLY/HOMEWORK

Homework Assignment

_____ Transitional: Day 1: SRH p. 769 1–15 odd; pp. 166–169 Exs. 20–23, 30–33, 50–57, 62–64
Day 2: pp. 167–169 Exs. 36, 39, 40, 43–49, Quiz 2

_____ Average: Day 1: pp. 166–169 Exs. 24–27, 32, 33, 50–70 even
Day 2: pp. 167–169 Exs. 37, 38, 41–49 Quiz 2

_____ Advanced: Day 1: pp. 166–169 Exs. 28, 29, 34, 35, 58–61, 65–70
Day 2: pp. 167–169 Exs. 36–38, 41–53, Quiz 2, EC: CRB p. 82

Reteaching the Lesson

_____ Practice Masters: CRB pages 76–77 (Level A, Level B)

_____ Reteaching with Practice: CRB pages 78–79 or Practice Workbook with Examples; Resources in Spanish

_____ Personal Student Tutor: CD-ROM

Extending the Lesson

_____ Interdisciplinary/Real-Life Applications: CRB page 81

_____ Challenge: CRB page 82

ASSESSMENT OPTIONS

_____ Daily Quiz (3.6): TE page 169, CRB page 86, or Transparencies

_____ Standardized Test Practice: SE page 168; STP Workbook; Transparencies

_____ Quiz 3.4–3.6: SE page 169; CRB page 83; Resources in Spanish

Notes _____

_____

TEACHER'S NAME _____ CLASS _____ ROOM _____ DATE _____

Lesson Plan for Block Scheduling

1-block lesson (See *Pacing the Chapter,* TE page 128A)   For use with pages 163–170

GOAL   **Find exact and approximate solutions of equations that contain decimals.**

State/Local Objectives _____

_____

✓ **Check the items you wish to use for this lesson.**

STARTING OPTIONS

____ Homework Check (3.5): TE page 160; Answer Transparencies
____ Homework Quiz (3.5): TE page 162,
      CRB page 72, or Transparencies
____ Warm-Up: CRB page 72 or Transparencies

TEACHING OPTIONS

____ Lesson Opener: CRB page 73 or Transparencies
____ Examples Day 1: 1–2, SE pages 163–164; Day 2: 3–4, SE pages 164–165
____ Extra Examples: TE pages 164–165 or Transparencies; Internet Help at *www.mcdougallittell.com*
____ Checkpoint Exercises: Day 1: Exs. 1–7, SE pages 163–164; Day 2: Ex. 8, SE page 165
____ Technology Graphing Calculator: Solving Multi-Step Equations: SE page 170;
      CRB page 75 (Keystrokes)
____ Graphing Calculator Activity with Keystrokes: CRB pages 74–75
____ Concept Check: TE page 165
____ Guided Practice Exercises: SE page 166; Day 1: Exs. 1–18; Day 2: Ex. 19

APPLY/HOMEWORK

Homework Assignment (See also the assignment for Lessons 3.5 and 3.7.)

____ Block Schedule: Day 1: pp. 166–169 Exs. 24–27, 32, 33. 50–70 even
      Day 2: pp. 167–169 Exs. 37, 38, 41–49, Quiz 2

Reteaching the Lesson

____ Practice Masters: CRB pages 76–77 (Level A, Level B)
____ Reteaching with Practice: CRB pages 78–79 or Practice Workbook with Examples;
      Resources in Spanish
____ Personal Student Tutor: CD-ROM

Extending the Lesson

____ Interdisciplinary/Real-Life Applications: CRB page 81
____ Challenge: CRB page 82

ASSESSMENT OPTIONS

____ Daily Quiz (3.6): TE page 169, CRB page 86, or Transparencies
____ Standardized Test Practice: SE page 168; STP Workbook; Transparencies
____ Quiz 3.4–3.6: SE page 169; CRB page 83; Resources in Spanish

Notes _____

CHAPTER PACING GUIDE	
Day	**Lesson**
1	3.1 (all)
2	3.2 (all); 3.3 (all)
3	3.4 (all)
4	3.5 (all); **3.6 (begin)**
5	**3.6 (end);** 3.7 (all)
6	3.8 (all); 3.9 (all)
7	Ch. 3 Review and Assess

Lesson 3.6

NAME _____ DATE _____

WARM-UP EXERCISES

For use before Lesson 3.6, pages 163–170

Lesson 3.6

Round each number to the indicated place value.

1. 1041; tens

2. 14.256; hundredths

3. 22.0401; tenths

4. -14.691; ones

5. -3.755; hundredths

....

DAILY HOMEWORK QUIZ

For use after Lesson 3.5, pages 157–162

Solve the equation.

1. $2(x - 3) = 5(x + 3)$

2. $6(y + 3) = 6 + 8(y - 1)$

3. $7a - 5(8 - a) = 2(a + 5)$

4. $\frac{1}{4}(12x - 4) = -3(-3 + x) - 5x$

5. A swimming pool charges $3 per session to swim, without a discount card. A discount card costs $36 and with the card it costs only $1 per session to swim. How many times would you have to swim to justify buying the discount card?

Algebra 1
Chapter 3 Resource Book

NAME _____ DATE _____

Calculator Lesson Opener

For use with pages 163–170

1. Answer the questions to solve $-18x + 12 = 116$. Use a calculator.

   a. What is the first step in solving this equation? Use your calculator to complete this step.

   b. What is the second step in solving this equation? Do you need to re-enter the result of your first step to complete the second step? Explain. Use your calculator to complete the second step.

   c. Is the solution shown on your calculator an exact solution? Why or why not?

2. Answer the questions to solve $6.6x + 4.5 = 0.9x - 1.2$. Use a calculator.

   a. What is the first step to solve this equation? Use your calculator to complete this step. Store the result in memory.

   b. What is the second step to solve this equation? Use your calculator to complete the second step.

   c. What is the third step to solve this equation? Explain how you can use the amount you stored in memory in part (a) to complete the third step. Then use your calculator to complete this step.

   d. Is the solution shown on your calculator an exact solution? Why or why not?

Explain how to use a calculator to solve the equation. Then solve the equation.

3. $12x + 23 = 98$

4. $2.4x - 1.5 = -9.8$

5. $10x - 126 = 2x + 45$

6. $-1.9x - 2.8 = 0.5x - 0.5$

Graphing Calculator Activity

For use with page 170

Keystrokes for Excel

Open computer to Excel program.

Select cell A1.

X `TAB` Y1 `TAB` Y2

Enter *x*-values 1–16 in cells A2–A17.

Select cell B2.

= 4.29 * A2 + 3.89 * (8 − A2) `ENTER`

Select cell B2. From the **Edit** menu, choose **Copy.**

Select cells B3–B17. From the **Edit** menu, choose **Paste.**

Select cell C2.

= 2.65 * A2 `ENTER`

Select cell C2. From the **Edit** menu, choose **Copy.**

Select cells C3–C17. From the **Edit** menu, choose **Paste.**

Select cell A19.

Enter *x*-values 13.1–13.9 in cells A19–A27.

Select cell B2. From the **Edit** menu, choose **Copy.**

Select cells B19–B27. From the **Edit** menu, choose **Paste.**

Select cell C2.

Select cells C19–C27. From the **Edit** menu, choose **Paste.**

Graphing Calculator Activity Keystrokes

For use with page 170

Keystrokes for Technology Activity 3.6

TI-83

| 2nd | WINDOW | 0 | ENTER | 1 | ENTER |

| Y= | 4.29 | X,T,θ,*n* | + | 3.89 | (| 8 − | X,T,θ,*n* |) | ENTER |

2.65 | X,T,θ,*n* | ENTER |

| 2nd | GRAPH |

Use [▼] key to scroll down until L_1 contains 13, 14, and 15 in the viewing window.

| 2nd | WINDOW |

13.1 | ENTER | 0.1 | ENTER |

| 2nd | GRAPH |

Use [▼] key to scroll down until L_1 contains 13.8, 13.9, and 14.0 in the viewing window.

NAME _____ DATE _____

Practice A

For use with pages 163–170

Perform any indicated operation. Round the result to the nearest tenth and then to the nearest hundredth.

1. 42.8451

2. -3.0624

3. $1.0847 + 62.5583$

4. $24.0321 - 21.8217$

5. $-23.981(-4.598)$

6. $15.953 \div 3.476$

Solve the equation. Round the result to the nearest hundredth. Check the rounded solution.

7. $3x + 18 = 26$

8. $7x - 1 = 8$

9. $-2 + 4x = 13$

10. $17 = 18 - 6x$

11. $8x - 3 = -24$

12. $39 = -3x + 2$

13. $-5x + 21 = 80$

14. $21x + 3 = 121$

15. $14(3 - 7x) = 9$

Solve the equation. Round the result to the nearest hundredth.

16. $2.3x + 4.8 = 9.3$

17. $5.1x - 7.2 = 1.4$

18. $-6.4x + 7.8 = 8.8$

19. $1.85 = 3.02 + 2.51x$

20. $-5.89x + 7.5 = 2.18$

21. $2.38x + 6.8 = 3.94x - 3.44$

22. $2(0.78 + 0.04x) = 0.01x$

23. $2.7 - 3.6x = 8.4 + 23.7x$

24. $5.3 + 9.2x = 7.4x - 8.8$

25. *Cutting a Rope* A rope which is 10.3 centimeters long is cut into 4 pieces of equal length. What is the length of each piece?

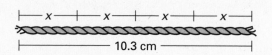

26. *Saving Money* You are saving money for the purchase of a new bike. You have saved $78.23. The bike costs $152.95. How much more money do you need?

| Money saved | + | Money needed | = | Price of bike |

27. *Track Meet* The winner of the track meet had an average speed of 5 meters per second. The second place runner had an average speed of 4.5 meters per second. If the winner finished 2.2 seconds ahead of the second place runner, how long did it take the winner to cross the finish line?

| Rate of winner | · | Time of winner (t) | = | Rate of 2nd place runner | · | Time of 2nd place runner ($t + 2.2$) |

Algebra 1
Chapter 3 Resource Book

Practice B

For use with pages 163–170

Perform any indicated operation. Round the result to the nearest tenth and then to the nearest hundredth.

1. -42.8451

2. $-3.06(5.98)$

3. $61.0847 + 62.5583$

4. $-24.0321 - 21.8219$

5. $-23.981(-4.525)$

6. $15.951 \div 3.476$

Solve the equation. Round the result to the nearest hundredth. Check the rounded solution.

7. $3x + 13 = 26$

8. $7x - 9 = 16$

9. $-2 + 4x = -13$

10. $17 = 48 - 6x$

11. $8x - 3 = -54$

12. $-67 = -14x + 29$

13. $-19x - 21 = -80$

14. $21x - 33 = -121$

15. $12(2x - 11) = 3x + 43$

16. $-x = 5(-6 - 7x)$

17. $2(x - 3) = 5x + 7$

18. $14(3 - 7x) = -(9 + 2x)$

Solve the equation. Round the result to the nearest hundredth.

19. $2.39x + 4.82 = 9.37 + 6.55x$

20. $35.13x - 7.26 = 11.48 - 14.91x$

21. $2.75 - 23.68x = 8.04 + 23.17x$

22. $51.56 - 29.28x = -47.78x - 68.82$

23. $-7.41x - 3.69 = 9.82$

24. $5.28x - 3.96 = 2.38x - 1.59$

25. $18.27 - 13.23x = -9.5x - 12.4$

26. $3.44 - 1.25x = 2.5x + 6.81$

Multiply the equation by a power of 10 to write an equivalent equation with integer coefficients. Solve the equation. Round the result to the nearest hundredth.

27. $-1.8 + 4.1x = 5.7$

28. $5.3 + 9.2x = 7.4x - 8.8$

29. $26.4x - 3.2 = 5.9x - 32.1$

30. *Purchase of Bike* You are shopping for a new bike. The sales tax is 6%. You have a total of $159 to spend. What is your price limit for the bike?

$$\boxed{\begin{array}{c}\text{Price}\\\text{limit}\end{array}} + \boxed{\begin{array}{c}\text{Sales}\\\text{tax rate}\end{array}} \cdot \boxed{\begin{array}{c}\text{Price}\\\text{limit}\end{array}} = \boxed{\begin{array}{c}\text{Total}\\\text{cost}\end{array}}$$

31. *Track Meet* The winner of the track meet had an average speed of 5 meters per second. The second place runner had an average speed of 4.5 meters per second. If the winner finished 1.2 seconds ahead of the second place runner, how long did it take the winner to cross the finish line?

$$\boxed{\begin{array}{c}\text{Rate of}\\\text{winner}\end{array}} \cdot \boxed{\begin{array}{c}\text{Time of}\\\text{winner}\end{array}} = \boxed{\begin{array}{c}\text{Rate of 2nd}\\\text{place runner}\end{array}} \cdot \boxed{\begin{array}{c}\text{Time of 2nd}\\\text{place runner}\end{array}}$$

32. *Buying a Sweatshirt* You have $32.14 to spend for a sweatshirt. The sales tax is 5%. What is the most the sweatshirt can cost?

Reteaching with Practice

For use with pages 163–170

GOAL Find exact and approximate solutions of equations that contain decimals.

> **VOCABULARY**
>
> A **rounding error** occurs when you use solutions that are not exact.

EXAMPLE 1 *Rounding for the Final Answer*

Solve $412x - 1640 = 238x - 12$. Round the result to the nearest hundredth.

SOLUTION

$$412x - 1640 = 238x - 12 \qquad \text{Write original equation.}$$
$$174x - 1640 = -12 \qquad \text{Subtract } 238x \text{ from each side.}$$
$$174x = 1628 \qquad \text{Add 1640 to each side.}$$
$$x = \frac{1628}{174} \qquad \text{Divide each side by 174.}$$
$$x \approx 9.35632 \qquad \text{Use a calculator.}$$
$$x \approx 9.36 \qquad \text{Round to the nearest hundredth.}$$

The solution is approximately 9.36.

Exercises for Example 1
....

Solve the equation. Round the result to the nearest hundredth.

1. $21x + 60 = 72$
2. $23 - 5x = 114x + 30$
3. $44y - 18 = y + 17$
4. $12b - 93 = 54b - 142$

EXAMPLE 2 *Solving an Equation Containing Decimals*

Solve $3.11x - 11.75 = 2.02x$. Round to the nearest hundredth.

SOLUTION

$$3.11x - 11.75 = 2.02x \qquad \text{Write original equation.}$$
$$1.09x - 11.75 = 0 \qquad \text{Subtract } 2.02x \text{ from each side.}$$
$$1.09x = 11.75 \qquad \text{Add 11.75 to each side.}$$
$$x = \tfrac{11.75}{1.09} \qquad \text{Divide each side by 1.09.}$$
$$x \approx 10.77981 \qquad \text{Use a calculator.}$$
$$x \approx 10.78 \qquad \text{Round to nearest hundredth.}$$

The solution is approximately 10.78. Check this in the original equation.

Reteaching with Practice

For use with pages 163–170

Exercises for Example 2

Solve the equation and round to the nearest hundredth.

5. $22.5 + 3.2x = 3.4x$

6. $-0.83y + 0.17 = 0.72y$

EXAMPLE 3 *Using a Verbal Model*

While dining at a restaurant, you want to leave a 15% tip. You have a total of $14.00 to spend. What is your price limit for the dinner? Using the verbal model below, write and solve an algebraic equation.

$$\boxed{\text{Price limit}} + \boxed{\text{Tip rate}} \cdot \boxed{\text{Price limit}} = \boxed{\text{Total cost}}$$

SOLUTION

Let x represent your price limit.

$x + 0.15x = 14.00$	Write algebraic model.
$1.15x = 14.00$	Combine like terms.
$x = \dfrac{14.00}{1.15}$	Divide each side by 1.15.
$x \approx 12.173913$	Use a calculator.
$x \approx 12.17$	Round down.

The answer is rounded *down* to $12.17 because you have a limited amount to spend.

Exercises for Example 3

7. Rework Example 3 if you have $16.00 to spend.

8. Rework Example 3 if you want to leave a 20% tip.

NAME _____ DATE _____

Quick Catch-Up for Absent Students

For use with pages 163–170

The items checked below were covered in class on (date missed) _____

Lesson 3.6: Solving Decimal Equations

_____ **Goal:** Find exact and approximate solutions of equations that contain decimals.

Material Covered:

_____ Example 1: Round for the Final Answer

_____ Example 2: Solve an Equation that Contains Decimals

_____ Example 3: Round for a Practical Answer

_____ Example 4: Use a Verbal Model

Technology Activity: Solving Multi-Step Equations (p. 170)

_____ **Goal:** Use a graphing calculator to solve multi-step equations.

Vocabulary:

rounding error, p. 164

_____ Other (specify)_____

_____

Homework and Additional Learning Support

_____ Textbook (specify) pp. 166–169 _____

_____

_____ Internet: Extra examples at www.mcdougallittell.com

_____ *Reteaching with Practice* worksheet (specify exercises) _____

_____ *Personal Student Tutor* for Lesson 3.6

NAME _____ DATE _____

Interdisciplinary Application

For use with pages 163–170

Swimming

HISTORY   In 1896, the modern Olympics were first held. Among the competitions were several men's swimming events, all using the breaststroke. The 1900 Olympic games added backstroke events. And in the 1912 games, women's Olympic swimming competition began.

During the course of the twentieth century, the trend shows that the records and the ages of leading swimmers are both decreasing.

In 1996, Alexandar Popov of Russia won the Olympic medal in the men's 100 yard freestyle event with a time of 48.74 seconds.

In Exercises 1–4, use the information above.

1. The Olympic gold medal winning 100 meter freestyle times for men have been decreasing at a rate of about 0.335 second per year. Write an expression for the men's gold medal winning time x years after 1996.

2. In 1996, Le Jingyi won the Olympic gold medal in the women's freestyle event with a time of 54.5 seconds. The Olympic gold medal winning 100 meter freestyle times for women have been decreasing at a rate of about 0.37 second per year. Write an expression for the women's gold medal winning time x years after 1996.

3. Write and solve an equation to find the number of years it will take after 1996 for the men's and women's Olympic gold medal winning 100 meter freestyle times to be the same.

4. What will the record time be when the men's and women's times are the same? What does this tell you?

NAME _____ DATE _____

Challenge: Skills and Applications

For use with pages 163–170

1. Find the population density for each country in the table.
   Round to the nearest tenth of a person per square mile.

Country	Population (in millions)	Area (in hundred thousand square miles)	Population density (in people per square mile)
Germany	84.1	1.38	
India	967.6	12.22	
Japan	125.7	1.46	
Turkey	63.5	3.01	
United States	268.0	36.75	

2. About how many times more crowded is the most crowded
   country in the table than the least crowded one?

3. If a person is chosen randomly from the population of the
   5 countries in the table, what is the probability the person
   lives in Germany? Round to the nearest thousandth.

For Exercises 4–8, use the following information.

A satellite is in orbit around Earth, travelling at a rate of
85.4 minutes per orbit. A space shuttle enters the same orbit,
travelling at a rate of 93.6 minutes per orbit. The shuttle is
0.25 orbits behind the satellite, travelling toward it.

4. Let x be the number of orbits the shuttle makes before the
   shuttle and the satellite first meet. Which expression below
   might represent the number of orbits the satellite makes in the
   same time? (*Hint:* Think about the starting positions and
   about which vehicle makes more orbits in a given time.)

   A. $x - 0.25$ **B.** $x + 0.25$ **C.** $x - 0.75$ **D.** $x + 0.75$

5. Use your answer to Exercise 4 to write expressions for the
   time the shuttle travels and for the time the satellite travels
   before they first meet. Then write an equation that relates
   the two expressions.

6. Solve the equation from Exercise 5. Round to the nearest
   hundredth. About how many orbits does each spacecraft
   make before they first meet?

7. Explain how you can use the minutes per orbit of the two
   spacecrafts to find the number of orbits the satellite makes
   as the shuttle makes n orbits, for any value of n. Use this
   method to check your answer to Exercise 6.

8. How many hours and minutes will it take before the satellite and the shuttle first meet? Round to the nearest whole
   minute.

NAME _____ DATE _____

Quiz 2

For use after Lessons 3.4–3.6

Solve the equation. (Lesson 3.4)

1. $11p + 2 = 2(p - 8)$

2. $3(6 + 9t) = 45t - 180$

3. $7 - 15s = 5s - 3$

4. You are planning on buying a bike and are offered two payment plans. Plan A requires a $120 down payment and x dollars a month for 4 months. In Plan B, there is no down payment and the payment is $10 more per month than in Plan A for 8 months. What is the monthly payment under each plan? *(Lesson 3.5)*

   a. Write an equation to model the problem.

   b. Solve the equation and answer the question.

Solve the equation. Round the result to the nearest hundredth. (Lesson 3.6)

5. $5.61 - 8.94m = 3.76m + 4.86$

6. $9.66 + 2.73 = 10.68y - 18.32$

Answers

1. _____

2. _____

3. _____

4. _____

5. _____

6. _____

TEACHER'S NAME _____ CLASS _____ ROOM _____ DATE _____

Lesson Plan

1-day lesson (See *Pacing the Chapter,* TE page 128A)   **For use with pages 171–176**

GOAL Solve a formula for one of its variables.

State/Local Objectives _____

_____

✓ Check the items you wish to use for this lesson.

STARTING OPTIONS
____ Homework Check (3.6): TE page 166; Answer Transparencies
____ Homework Quiz (3.6): TE page 169, CRB page 86, or Transparencies
____ Warm-Up: CRB page 86 or Transparencies

TEACHING OPTIONS
____ Lesson Opener: CRB page 87 or Transparencies
____ Examples 1–5: SE pages 171–173
____ Extra Examples: TE pages 172–173 or Transparencies; Internet Help at *www.mcdougallittell.com*
____ Checkpoint Exercises: SE pages 172–173
____ Concept Check: TE page 173
____ Guided Practice Exercises: SE page 174

APPLY/HOMEWORK
Homework Assignment
____ Transitional: EP: p. 160 Exs. 25–30, pp. 174–176 Exs. 11–14, 20, 21, 25, 26, 27–45 odd
____ Average: pp. 174–176 Exs. 15, 16, 18–21, 24–26, 30–32, 36–42, 43–46
____ Advanced: pp. 174–176 Exs. 15–19, 22–26, 28–46 even; EC: CRB p. 94

Reteaching the Lesson
____ Practice Masters: CRB pages 88–89 (Level A, Level B)
____ Reteaching with Practice: CRB pages 90–91 or Practice Workbook with Examples;
Resources in Spanish
____ Personal Student Tutor: CD-ROM

Extending the Lesson
____ Interdisciplinary/Real-Life Applications: CRB page 93
____ Challenge: CRB page 94

ASSESSMENT OPTIONS
____ Daily Quiz (3.7): TE page 176, CRB page 97, or Transparencies
____ Standardized Test Practice: SE page 176; STP Workbook; Transparencies

Notes _____

_____

_____

_____

TEACHER'S NAME _____ CLASS _____ ROOM _____ DATE _____

Lesson Plan for Block Scheduling

Half-block lesson (See *Pacing the Chapter,* TE page 128A)   For use with pages 171–176

GOAL   Solve a formula for one of its variables.

State/Local Objectives _____

_____

_____

✓ **Check the items you wish to use for this lesson.**

STARTING OPTIONS

____ Homework Check (3.6): TE page 166; Answer Transparencies
____ Homework Quiz (3.6): TE page 169,
        CRB page 86, or Transparencies
____ Warm-Up: CRB page 86 or Transparencies

TEACHING OPTIONS

____ Lesson Opener: CRB page 87 or Transparencies
____ Examples 1–5: SE pages 171–173
____ Extra Examples: TE pages 172–173 or Transparencies; Internet Help at *www.mcdougallittell.com*
____ Checkpoint Exercises: SE pages 172–173
____ Concept Check: TE page 173
____ Guided Practice Exercises: SE page 174

APPLY/HOMEWORK

Homework Assignment (See also the assignment for Lesson 3.6.)
____ Block Schedule: pp. 174–176 Exs. 15, 16, 18–21, 24–26, 32–38, 43–46

Reteaching the Lesson

____ Practice Masters: CRB pages 88–89 (Level A, Level B)
____ Reteaching with Practice: CRB pages 90–91 or Practice Workbook with Examples;
        Resources in Spanish
____ Personal Student Tutor: CD-ROM

Extending the Lesson

____ Interdisciplinary/Real-Life Applications: CRB page 93
____ Challenge: CRB page 94

ASSESSMENT OPTIONS

____ Daily Quiz (3.7): TE page 176, CRB page 97, or Transparencies
____ Standardized Test Practice: SE page 176; STP Workbook; Transparencies

Notes _____

_____

_____

_____

| CHAPTER PACING GUIDE ||
Day	Lesson
1	3.1 (all)
2	3.2 (all); 3.3 (all)
3	3.4 (all)
4	3.5 (all); 3.6 (begin)
5	3.6 (end); **3.7 (all)**
6	3.8 (all); 3.9 (all)
7	Ch. 3 Review and Assess

Lesson 3.7

Algebra 1
Chapter 3 Resource Book

NAME _____ DATE _____

WARM-UP EXERCISES

For use before Lesson 3.7, pages 171–176

Find the value of each expression when $a = 3$ and $b = -4$.

1. $a - b$

2. $2ab$

3. $6a + 12a - b$

4. $b(5 - a)$

5. $(-3b)(-b)$

DAILY HOMEWORK QUIZ

For use after Lesson 3.6, pages 163–170

Solve the equation. Round the result to the nearest hundredth.

1. $14x + 6 = 31$

2. $-48m + 56 = 123$

3. $5.23x = 21.47 - 6.91x$

4. $13.19 - 8.54x = 7.94x - 1.82$

5. The student council sold pencils as a fundraiser and made $378.90. If there are 16 student council members, about how much did each earn in sales for the council?

Algebra 1
Chapter 3 Resource Book

Application Lesson Opener

For use with pages 171-176

Work in a group.

1. Write at least four formulas you know. If you do not remember the exact formula using variables, describe the formula in words. Then try to write it using variables. Try to choose at least one formula from geometry, one from science, one related to consumer mathematics, and one related to sports or hobbies.

2. Choose one of your formulas and explain what the variables represent.

3. Give an example of how each formula is used in everyday life.

4. Share your formulas with your classmates. Ask them if they can think of other everyday uses of your formula.

Lesson 3.7

NAME _____ DATE _____

Practice A

For use with pages 171–176

Solve for the indicated variable.

1. *Area of a Rectangle*
   Solve for w: $A = \ell w$

2. *Circumference of a Circle*
   Solve for r: $C = 2\pi r$

3. *Perimeter of a Square*
   Solve for s: $P = 4s$

4. *Distance*
   Solve for r: $d = rt$

5. *Ohm's Law*
   Solve for R: $E = IR$

6. *Volume of a Circular Cone*
   Solve for h: $V = \dfrac{1}{3}\pi r^2 h$

Solve the equation for *y*.

7. $-3x + y = 8$

8. $y - 2x = 15$

9. $y + 7x = 4$

10. $6x + 3y = -12$

11. $-x = y - 12$

12. $4y - 8 = 12x$

13. $2 - y = 8x$

14. $\dfrac{1}{2}y - 7 = 3x$

15. $2x + 3y - 6 = 9$

16. $-2x + 5y - 7 = -12$

17. $4x + 2y = 8x - 5$

18. $2(y + 5) = 4x$

Solve the equation for *x*. Then use the result to find *x* when *y* = −2, −1, 0, and 1.

19. $x - y = 4$

20. $x - 2y = -3$

21. $2x + 4y - 6 = 0$

Airplane Travel   **In Exercises 22 and 23, use the formula $d = rt$, where d is the distance traveled at a rate of r for time t.**

22. Solve the equation for t.

23. Determine how long it will take an airplane to travel 2000 miles if it flies 200 miles per hour, 400 miles per hour, and 600 miles per hour.

Savings Account   **In Exercises 24 and 25, use the formula $I = Prt$, where I is the simple interest on an investment of P dollars at an interest rate r for t years.**

24. Solve the equation for r.

25. Find the interest rate r for an investment of $1200 that earned $66 in interest for one year.

Solve for the indicated variable.

1. *Simple Interest*
Solve for t: $I = Prt$

2. *Area of a Kite*
Solve for d_2: $A = \frac{1}{2}d_1 d_2$

3. *Area of a Trapezoid*
Solve for b_1: $A = \frac{1}{2}h(b_1 + b_2)$

4. *Temperature*
Solve for C: $F = \frac{9}{5}C + 32$

5. *Surface Area of a Regular Pyramid*
Solve for ℓ: $S = B + \frac{1}{2}P\ell$

6. *Surface Area of a Right Cylinder*
Solve for h: $S = 2\pi r^2 + 2\pi rh$

Solve the equation for *y*.

7. $y + 9x = 4$

8. $5y - 2x = 15$

9. $-2y + 10x = 8$

10. $-4y - 8 = 12x$

11. $-4x = 2y - 16$

12. $7 - y = 3.5x$

13. $2 - \frac{y}{6} = 8x$

14. $\frac{1}{2}y - 7 = -3x$

15. $3y - 6 = 9 - 2x$

16. $-2x + 5y - 6 = -11$

17. $4x - 8x + 4 = 2y - 5$

18. $\frac{1}{2}(y + 5) + 6x = 4x$

19. $3x + 3y = 14 - 4x$

20. $4x + 2(y - 3) = 10$

21. $6x - 3(y - 1) = 4x + 8$

Solve the equation for *x*. Then use the result to find *x* when *y* = −2, −1, 0, and 1.

22. $x - 2y = -3$

23. $5x - y = 10$

24. $4x - 2y = 4$

25. $5y + x = -3 + 4y$

26. $3x - 2y = 4 + 7x$

27. $4y - 3(x - 2) = 22$

Airplane Travel **In Exercises 28 and 29, use the formula** $d = rt$**, where** *d* **is the distance traveled at a rate of** *r* **for time** *t*.

28. Solve the equation for t.

29. Determine how long (in hours and minutes) it will take an airplane to travel 2500 miles if it flies 200 miles per hour, 400 miles per hour, and 600 miles per hour.

Savings Account **In Exercises 30 and 31, use the formula** *I* = *Prt*, **where** *I* **is the simple interest on an investment of** *P* **dollars at an interest rate** *r* **for** *t* **years.**

30. Solve the equation for P.

31. Find the principal P for two years that earned $151.25 in interest with a rate of 0.055.

Lesson 3.7

NAME _____ DATE _____

Reteaching with Practice

For use with pages 171–176

GOAL Solve a formula for one of its variables

VOCABULARY

A **formula** is an algebraic equation that relates two or more real-life quantities.

EXAMPLE 1 *Solving and Using an Area Formula*

Use the formula for the area of a rectangle, $A = \ell w$.

a. Solve the formula for the width w.

b. Use the new formula to find the width of a rectangle that has an area of 72 square inches and a length of 9 inches.

SOLUTION

a. Solve for width w.

$A = \ell w$ Write original formula.

$\dfrac{A}{\ell} = \dfrac{\ell w}{\ell}$ To isolate w, divide each side by ℓ.

$\dfrac{A}{\ell} = w$ Simplify.

b. Substitute the given values into the new formula.

$w = \dfrac{A}{\ell} = \dfrac{72}{9} = 8$

The width of the rectangle is 8 inches.

Exercises for Example 1
....
Solve for the indicated variable.

1. Area of a Triangle

Solve for h: $A = \frac{1}{2}bh$

2. Circumference of a Circle

Solve for r: $C = 2\pi r$

3. Simple Interest
Solve for P: $I = Prt$

4. Simple Interest
Solve for r: $I = Prt$

Reteaching with Practice

For use with pages 171–176

EXAMPLE 2 *Solving and Using a Distance Formula*

Driving on the highway, you travel 930 miles at an average speed of 62 miles per hour.

a. Solve the distance formula $d = rt$ for time t.

b. Estimate the time spent driving.

SOLUTION

a. $d = rt$ Write original formula.

$\dfrac{d}{r} = t$ Divide each side by r.

b. Substitute the given values for the new formula.

$t = \dfrac{d}{r}$ Write formula.

$t = \dfrac{930}{62}$ Substitute 930 for d and 62 for r.

$t = 1.5$ Solve for t.

Answer You spent approximately 1.5 hours driving.

Exercises for Example 2

5. Solve the distance formula for r.

6. Use the result from Exercise 5 to find the average speed in miles per hour of a car that travels 1275 miles in 22 hours. Round your answer to the nearest whole number.

Lesson 3.7

NAME _____ DATE _____

Quick Catch-Up for Absent Students

For use with pages 171–176

The items checked below were covered in class on (date missed) _____

Lesson 3.7: Formulas

____ **Goal:** Solve a formula for one of its variables.

Material Covered:

   ____ Student Help: Look Back

   ____ Example 1: Solve a Temperature Conversion Formula

   ____ Example 2: Solve an Area Formula

   ____ Example 3: Solve and Use an Area Formula

   ____ Example 4: Solve and Use a Density Formula

   ____ Example 5: Solve and Use a Distance Formula

Vocabulary:

   formula, p. 171

____ Other (specify) _____

_____

Homework and Additional Learning Support

   ____ Textbook (specify) <u>pp. 174–176</u>_____

   _____

   ____ Internet: Extra example at www.mcdougallittell.com

   ____ *Reteaching with Practice* worksheet (specify exercises)_____

   ____ *Personal Student Tutor* for Lesson 3.7

NAME _____ DATE _____

Interdisciplinary Application

For use with pages 171–176

Magnification

OPTICS   Prisms spread or disperse beams of light. If you put one in your window, it will break the sunlight passing through it into a rainbow of color on your wall. Beams of light can also be bent by lenses, which are curved pieces of glass or plastic that can be found in cameras and eyeglasses. A common type of lens, the converging lens, brings parallel rays of light together.

Objects viewed through a lens often appear larger (or smaller) than they are. This process is called magnification. The magnification of a lens is usually measured in terms of the magnification power, which is simply the height of the image (h_i) divided by the height of the actual object (h_o).

$$M = \frac{h_i}{h_o}$$

If the heights of the object and its image are not given, you may also divide the distance from the image to the lens (d_i) by the distance of the lens to the object (d_o) to find the amount of magnification.

$$M = \frac{d_i}{d_o}$$

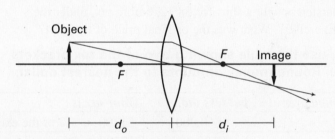

In Exercises 1-5, use the information above.

1. An object 3 inches high has an image 9 inches high. Find the amount of magnification.

2. An object is placed 6 centimeters in front of a lens and its image is 15 centimeters from the lens. Find the amount of magnification.

3. An object 6 inches high is magnified 3.5 times.

   a. Solve $M = \dfrac{h_i}{h_o}$ for h_i.

   b. Find the height of the image.

4. The image of an object is 9 centimeters from a lens. The magnification power is $\frac{3}{4}$.

   a. Solve $M = \dfrac{d_i}{d_o}$ for d_o.

   b. Find the distance from the lens to the actual object.

5. In Exercise 4, the magnification power is $\frac{3}{4}$. What does this tell you about the image?

Challenge: Skills and Applications
For use with pages 171–176

In Exercises 1–4, solve the equation for x in terms of the other variables.

1. $p - \frac{6}{7}x = 2q$

2. $\frac{2}{3}ab - \frac{1}{5}x = 2$

3. $\frac{5}{2}(h - x) = \frac{k}{4}$

4. $-\frac{5}{8}x + r = 1 - \frac{t}{2}$

In Exercises 5–8, use the following information.

A men's clothing store decides to discount men's shirts by 14% and then give each customer a coupon for a further $5 discount on each shirt.

5. Write an equation that gives the fully discounted price d a customer would pay for a shirt as a function of the original price p. Then rewrite the function so that p is a function of d.

6. Clara Marsten bought a shirt for $8.12, before tax. What was the original price of the shirt?

7. Suppose the store applied the $5 discount coupon first and then took 14% off the price for a shirt. Write an equation for the price d a customer would pay for a shirt as a function of the original price p. Then rewrite the function so that p is a function of d.

8. Suppose Clara Marsten bought a shirt for $8.12, before tax, under the discount plan in Exercise 7. What was the original price of the shirt?

In Exercises 9–11, use the table showing two of the tax brackets for a certain state. Round money amounts to the nearest dollar.

For an annual income over . . .	but less than . . .	Your tax is . . .
$30,000	$60,000	$1200 + 5% of the excess over $30,000
$60,000	$100,000	$2700 + 7% of the excess over $60,000

9. For each bracket shown, write an equation that gives the amount of tax y a person in that bracket must pay as a function of his or her annual income x.

10. Rewrite each function from Exercise 9 so x is a function of y.

11. How much income would a taxpayer have who paid $3050 in tax?

Algebra 1
Chapter 3 Resource Book

Lesson Plan

1-day lesson (See *Pacing the Chapter,* TE page 128A)   For use with pages 177–182

GOAL **Use ratios and rates to solve real-life problems.**

State/Local Objectives _____

_____

✓ Check the items you wish to use for this lesson.

STARTING OPTIONS
____ Homework Check (3.7): TE page 174; Answer Transparencies
____ Homework Quiz (3.7): TE page 176, CRB page 97, or Transparencies
____ Warm-Up: CRB page 97 or Transparencies

TEACHING OPTIONS
____ Lesson Opener: CRB page 98 or Transparencies
____ Examples 1–6: SE pages 177–179
____ Extra Examples: TE pages 178–179 or Transparencies; Internet Help at *www.mcdougallittell.com*
____ Checkpoint Exercises: SE pages 177–179
____ Concept Check: TE page 179
____ Guided Practice Exercises: SE page 180

APPLY/HOMEWORK
Homework Assignment
____ Transitional: pp. 180–182 Exs. 13–19 odd, 22, 23, 28–33, 40, 41, 44–47, 49–63 odd
____ Average: pp. 180–182 Exs. 12–18 even, 24, 25, 28–31, 34, 35, 44–48, 52–59
____ Advanced: pp. 180–182 Exs. 20, 21, 26, 27, 36, 37, 42–48, 50, 53–55, 60–63; EC: CRB p. 106

Reteaching the Lesson
____ Practice Masters: CRB pages 99–100 (Level A, Level B)
____ Reteaching with Practice: CRB pages 101–102 or Practice Workbook with Examples; Resources in Spanish
____ Personal Student Tutor: CD-ROM

Extending the Lesson
____ Learning Activity: CRB page 104
____ Interdisciplinary/Real-Life Applications: CRB page 105
____ Challenge: CRB page 106

ASSESSMENT OPTIONS
____ Daily Quiz (3.8): TE page 182, CRB page 109, or Transparencies
____ Standardized Test Practice: SE page 182; STP Workbook; Transparencies

Notes _____

_____

_____

TEACHER'S NAME _____ CLASS _____ ROOM _____ DATE _____

Lesson Plan for Block Scheduling

Half-block lesson (See *Pacing the Chapter,* TE page 128A)   For use with pages 177–182

GOAL   Use ratios and rates to solve real-life problems.

State/Local Objectives _____

_____

_____

✓ **Check the items you wish to use for this lesson.**

STARTING OPTIONS

____ Homework Check (3.7): TE page 174; Answer Transparencies

____ Homework Quiz (3.7): TE page 176, CRB page 97, or Transparencies

____ Warm-Up: CRB page 97 or Transparencies

TEACHING OPTIONS

____ Lesson Opener: CRB page 98 or Transparencies

____ Examples 1–6: SE pages 177–179

____ Extra Examples: TE pages 178–179 or Transparencies; Internet Help at *www.mcdougallittell.com*

____ Checkpoint Exercises: SE pages 177–179

____ Concept Check: TE page 179

____ Guided Practice Exercises: SE page 180

APPLY/HOMEWORK

Homework Assignment (See also the assignment for Lesson 3.9.)

____ Block Schedule: pp. 180–182 Exs. 12–18 even, 24, 25, 28–31, 34, 35, 44–48, 52–59

Reteaching the Lesson

____ Practice Masters: CRB pages 99–100 (Level A, Level B)

____ Reteaching with Practice: CRB pages 101–102 or Practice Workbook with Examples; Resources in Spanish

____ Personal Student Tutor: CD-ROM

Extending the Lesson

____ Learning Activity: CRB page 104

____ Interdisciplinary/Real-Life Applications: CRB page 105

____ Challenge: CRB page 106

ASSESSMENT OPTIONS

____ Daily Quiz (3.8): TE page 182, CRB page 109, or Transparencies

____ Standardized Test Practice: SE page 182; STP Workbook; Transparencies

Notes _____

_____

_____

CHAPTER PACING GUIDE	
Day	**Lesson**
1	3.1 (all)
2	3.2 (all); 3.3 (all)
3	3.4 (all)
4	3.5 (all); 3.6 (begin)
5	3.6 (end); 3.7 (all)
6	**3.8 (all)**; 3.9 (all)
7	Ch. 3 Review and Assess

NAME _____ DATE _____

WARM-UP EXERCISES

For use before Lesson 3.8, pages 177–182

Write each fraction as a decimal and as a percent.

1. $\dfrac{4}{5}$

2. $\dfrac{8}{4}$

3. $\dfrac{9}{8}$

Write each ratio as a fraction in simplest form.

4. 14 to 16

5. 25:35

DAILY HOMEWORK QUIZ

For use after Lesson 3.7, pages 171–176

1. Solve $F = ma$ for a.

2. Solve $3y - x = 5$ for y.

3. Use $b = \dfrac{2A}{h}$ to find the value of b when $A = 6$ and $h = 4$.

4. Use $l = \dfrac{P - 2w}{2}$ to find l when $P = 16$ and $w = 3$.

5. The altitude of a hiker t minutes after he starts to climb is given by the formula below, where A is the altitude of the hiker, in feet.

$A = 18t + 10{,}500$

Solve for t and then find the time it will take to climb to an altitude of 10,878 feet.

Application Lesson Opener

For use with pages 177–182

The table shows the number of Democrats and Republicans in the Senate for the 102nd through the 106th Congresses.

Congress	*102nd*	*103rd*	*104th*	*105th*	*106th*
Democrats	56	57	48	45	45
Republicans	44	43	52	55	55

1. Find the ratio of Republicans to Democrats for the 106th Congress. Name another Congress with this same ratio.

2. Find the ratio of Democrats to Republicans for the 104th Congress. Explain how you can use this ratio to find the ratio of Republicans to Democrats for the 104th Congress.

3. For each Congress listed, there were 100 senators. What percent of the senators in the 102nd Congress were Democrats? Republicans?

The table shows the number of students and teachers in United States public schools during the school years listed.

School year	*91–92*	*92–93*	*93–94*	*94–95*	*95–96*
Students (millions)	42.0	42.8	43.5	44.1	44.8
Teachers (millions)	3.1	3.1	3.2	3.3	3.4

4. Use your calculator to find the number of students per teacher during each of the school years listed. Round to the nearest student.

   a. 91–92   **b.** 92–93   **c.** 93–94   **d.** 94–95   **e.** 95–96

NAME _____ DATE _____

Practice A

For use with pages 177–182

Find the unit rate.

1. $2 for 5 cans of vegetables

2. $67.50 for 3 concert tickets

3. $2.39 for 10 apples

4. 14 cups of flour for 4 loaves of bread

5. 32 ounces for 4 servings

6. 144 bottles in 12 cartons

Find the average speed.

7. Fly 780 miles in 2 hours.

8. Drive 1440 miles in 3 days.

9. Walk 1.2 miles in $\frac{1}{4}$ hour.

10. Read 115 pages in 2.5 hours.

11. Swim 3 kilometers in 40 minutes.

12. Mow 6 acres in 4 hours.

13. *Concert Tickets*   A concert sold out in 6 hours. A total of 9000 tickets were sold for the concert. At what rate did the tickets sell?

14. *Snowfall*   The record for snowfall in a 24-hour period is 76 inches at Silver Lake, Colorado, on April 14–15, 1921. At what rate did the snow fall on that day? Write the answer as a mixed number.

15. *Ballooning*   It took 86 hours for Joe Kittinger to cross the Atlantic in his balloon, *Rosie O'Grady*, a distance of 3543 miles. What was the average speed of the balloon? Round to the nearest tenth.

16. *Building a Model Railroad*   On an N-gauge model train set, a refrigerator car is 4.5 inches long. An actual refrigerator car is 60 feet long. What is the ratio of the length of the actual refrigerator car to the length of the model refrigerator car?

In Exercises 17 and 18, use the information given.

You have recorded your car mileage and gasoline use for several weeks.

Week	1	2	3	4
Number of miles	320	195	240	275
Number of gallons	12.4	7.3	9.6	10.7

17. Find the average consumption in miles per gallon for the 4-week period.

18. Use your answer to Exercise 17 to estimate the number of miles you can drive on 15 gallons of gasoline.

Practice B

For use with pages 177–182

Find the unit rate.

1. $4 for 5 cans of vegetables

2. $65.25 for 3 concert tickets

3. $2.49 for 10 apples

4. $65.25 for working 9 hours

5. 27 ounces for 4.5 servings

6. $2.99 for $1\frac{1}{2}$ quarts of juice

Find the average speed.

7. Fly 1500 miles in 4 hours.

8. Hike 46 miles in 3 days.

9. Walk 1.2 miles in 15 minutes.

10. Mow 3 acres in 4 hours.

11. Type 129 words in 3 minutes.

12. Drive 78 kilometers in $\frac{3}{4}$ hour.

13. *Concert Tickets*   A concert sold out in 6 hours. A total of 9570 tickets were sold for the concert. At what rate did the tickets sell?

14. *Rainfall*   The record for rainfall in a 24-hour period is 73.62 inches in Cilaos, Réunion, Indian Ocean, on March 15–16, 1952. At what rate did the rain fall on that day? Round to the nearest hundredth.

15. *Helium Ballooning*   It took about 144.25 hours for Richard Abruzzo and Troy Bradley to cross the Atlantic from Bangor, ME to Ben Slimane, Morocco, a distance of 3318.2 miles. What was the average speed of the balloon? Round to the nearest whole number.

16. *Building a Model Railroad*   On an N-gauge model train set, a tank car is 3.75 inches long. An actual tank car is 50 feet long. What is the ratio of the length of the actual tank car to the length of the model tank car?

In Exercises 17 and 18, use the information given.

You have recorded your car mileage and gasoline use for several weeks.

Week	1	2	3	4
Number of miles	320	195	240	280
Number of gallons	12.5	7.4	9.6	10.8

17. Find the average consumption in miles per gallon for the 4-week period.

18. Use your answer to Exercise 17 to estimate the number of miles you can drive on 15 gallons of gasoline.

Reteaching with Practice

For use with pages 177–182

GOAL Use ratios and rates to solve real-life problems.

VOCABULARY

If a and b are two quantities measured in different units, then the **rate of a per b** is $\frac{a}{b}$.

A **unit rate** is a rate per one given unit.

EXAMPLE 1 *Finding a Ratio*

The team won 12 of its 15 games. Find the ratio of wins to losses.

SOLUTION

$$\text{Ratio} = \frac{\text{games won}}{\text{games lost}} = \frac{12 \text{ games}}{3 \text{ games}} = \frac{4}{1}$$

Exercises for Example 1

1. You answer correctly 48 of the 50 questions on a quiz. Find the ratio of correct answers to incorrect answers.

EXAMPLE 2 *Finding a Unit Rate*

You hike 24 miles in 2 days. What is your average speed in miles per day?

SOLUTION

$$\text{Ratio} = \frac{24 \text{ miles}}{2 \text{ days}} = \frac{12 \text{ miles}}{1 \text{ day}} = 12 \text{ mi/day}$$

Exercises for Example 2

Find the unit rate.

2. A car drives 120 miles in 3 hours.

3. You earn $55 for working 5 hours.

EXAMPLE 3 *Using a Rate*

You took a survey of your classmates and found that 9 of the 27 classmates have public library cards. Use your results to make a prediction for the 855 students enrolled in your school.

Reteaching with Practice

For use with pages 177–182

SOLUTION

You can answer the question by writing a ratio. Let n represent the number of students in your school that have public library cards.

$$\frac{\text{Library cards in sample}}{\text{Total students in sample}} = \frac{\text{Library cards in school}}{\text{Total students in school}}$$

$$\frac{9}{27} = \frac{n}{855} \qquad \text{Write equation.}$$

$$855 \cdot \frac{9}{27} = n \qquad \text{Multiply each side by 855.}$$

$$285 = n \qquad \text{Simplify.}$$

Of the 855 students enrolled in the school, about 285 will have a public library card.

Exercises for Example 3

4. Rework Example 3 if 6 of the 27 class-mates have public library cards.

5. Rework Example 3 if 930 students are enrolled in the school.

EXAMPLE 4 *Applying Unit Analysis*

While visiting Italy you want to exchange \$120 for liras. The rate of currency exchange is 1850 liras per United States dollar. How many liras will you receive?

SOLUTION

You can use unit analysis to write an equation.

$$\text{dollars} \cdot \frac{\text{liras}}{\text{dollars}} = \text{liras}$$

$$D \cdot \frac{1850}{1} = L \qquad \text{Write equation.}$$

$$120 \cdot \frac{1850}{1} = L \qquad \text{Substitute 120 for } D \text{ dollars.}$$

$$222{,}000 = L \qquad \text{Simplify.}$$

You will receive 222,000 liras.

Exercises for Example 4

Convert the currency using the given exchange rate.

6. Convert \$150 U.S. dollars to German marks. (\$1 U.S. is 1.8943 marks)

7. Convert \$200 U.S. dollars to Austrian schillings. (\$1 U.S. is 13.3272 schillings)

NAME _____ DATE _____

Quick Catch-Up for Absent Students

For use with pages 177–182

The items checked below were covered in class on (date missed) _____

Lesson 3.8: Ratios and Rates

____ **Goal:** Use ratios and rates to solve real-life problems.

Material Covered:

    ____ Student Help: Writing Algebra

    ____ Example 1: Find a Ratio

    ____ Example 2: Find a Unit Rate

    ____ Student Help: Study Tip

    ____ Example 3: Find a Rate

    ____ Example 4: Use Unit Analysis

    ____ Example 5: Use a Rate

    ____ Example 6: Apply Unit Analysis

Vocabulary:

    ratio, p. 177   unit rate, p. 177

    rate, p. 177   unit analysis, p. 178

____ Other (specify)_____

_____

Homework and Additional Learning Support

    ____ Textbook (specify) pp. 180–182 _____

    _____

    ____ Internet: Extra examples at www.mcdougallittell.com

    ____ *Reteaching with Practice* worksheet (specify exercises) _____

    ____ *Personal Student Tutor* for Lesson 3.8

NAME _____ DATE _____

Learning Activity

For use with pages 177–182

GOAL **To find rectangles whose length-to-width ratios are equal to the golden ratio**

Materials: Ruler, loose-leaf paper, pencil

Exploring the Golden Ratio

The golden ratio, approximately 1.618034, was used by the ancient Greeks in their construction and artwork. The golden ratio is still used in today's architecture and art. In this activity, your group will try to find examples of the golden ratio in rectangles. The Greeks felt that a rectangle was very pleasing to the human eye if the ratio of the length to the width equaled the golden ratio. This kind of rectangle is called a golden rectangle.

Instructions

1. Write an equation that relates the length and width of a golden rectangle to the golden ratio. Solve the equation for the width.

2. Measure the lengths and widths of ten rectangles you find in your classroom and at home.

3. Use the equation from Step 1 to find the width that would make the rectangle a golden rectangle.

Golden Rectangle

Analyzing the Results

1. Are any of the rectangles you measured golden rectangles? If not, determine a width that would make the rectangle a golden rectangle.

2. Why do you think the Greeks found a golden rectangle more pleasing to the eye than other rectangles?

3. Examples of the golden ratio can also be found in nature. Can you find any?

NAME _____ DATE _____

Real-Life Application: When Will I Ever Use This?

For use with pages 177–182

Skyscrapers

Most buildings used to be constructed with strong outside walls to support the weight of the entire building. Skyscrapers, however, are designed with an internal steel skeleton that supports the structure. This steel frame construction made it possible to build skyscrapers. The invention of the elevator by Elisha Otis in the mid-nineteenth century made the construction of skyscrapers possible, for people could avoid the long, impractical stair climbs. With a design that enables the building to sway in strong winds and the advancements in fireproofing, the skyscraper became part of the city skyline.

The nine-story Home Insurance Building in Chicago, built in 1885, is the first "official" skyscraper. A 1916 zoning law in New York forces skyscrapers to be built in a stepped, or pyramid, fashion to allow for more air flow and sunlight at street level. Often comparable to cities, with their own shops, restaurants, banks, post offices, grocery stores, and apartments, skyscrapers are definitely one of architecture's crowning achievements. The top eight tallest buildings in the world in 1999 are listed in the table below.

Name	Petronas Tower 1	Petronas Tower 2	Sears Tower	Jin Mao Building	World Trade Center, 1
Height (feet)	1483	1483	1450	1380	1368
City	Kuala Lumpur	Kuala Lumpur	Chicago	Shanghai	New York
Built	1998	1998	1974	1999	1972

Name	World Trade Center, 2	Empire State	Central Plaza
Height (feet)	1362	1250	1227
City	New York	New York	Hong Kong
Built	1973	1931	1992

1. The Petronas Towers are 88 stories tall. What is the average number of feet per story?

2. When the Sears Tower was built, it was predicted to house 13,000 employees. Presently, 10,000 people work there during the day. What is the ratio of predicted employees to present employees?

3. The Jin Mao Building is 88 stories high. The top 38 levels are a hotel. What is the ratio of the hotel levels to stories in the building?

4. Over 40,000 people work in the World Trade Center and over 150,000 business and leisure visitors come each day. What is the ratio of workers to visitors?

Challenge: Skills and Applications

For use with pages 177–182

In Exercises 1–7, convert the measure. Round your answer to the nearest tenth.

1. 40 ft/sec to mi/h (1 mi = 5280 ft)

2. 65 lb/ft^3 to kg/L (1 kg $\approx$ 2.205 lb; 1 ft^3 $\approx$ 28.3 L)

3. 0.75 in./h to mm/min (1 in. = 25.4 mm)

4. 3500 in./min to mi/min

5. 0.28 lb/in. to lb/ft

6. 970 ft/sec to ft/min

7. 0.05 m/h to cm/h

8. While driving down the Aegean coast you see a sign that says gasoline costs 229,704 Turkish Lira (TL) per liter (L). The current exchange rate is 410,000 TL to a United States dollar. How much does the gasoline cost in dollars per gallon? (1 gal $\approx$ 3.784 L) Round to the nearest cent.

Lesson Plan

1-day lesson (See *Pacing the Chapter,* TE page 128A)   **For use with pages 183-188**

GOAL   **Solve percent problems.**

State/Local Objectives _____

_____

✓ Check the items you wish to use for this lesson.

STARTING OPTIONS

_____ Homework Check (3.8): TE page 180; Answer Transparencies

_____ Homework Quiz (3.8): TE page 182, CRB page 109, or Transparencies

_____ Warm-Up: CRB page 109 or Transparencies

TEACHING OPTIONS

_____ Lesson Opener: CRB page 110 or Transparencies

_____ Examples 1–4: SE pages 183–185

_____ Extra Examples: TE pages 184–185 or Transparencies; Internet Help at *www.mcdougallittell.com*

_____ Checkpoint Exercises: SE pages 184–185

_____ Concept Check: TE page 185

_____ Guided Practice Exercises: SE page 186

APPLY/HOMEWORK

Homework Assignment

_____ Transitional: pp. 186–188 Exs. 13–17, 22, 23, 28, 29, 35, 42–51, Quiz 3

_____ Average: pp. 186–188 Exs. 18, 19, 24, 25, 30, 31, 34–36, 40–49, Quiz 3

_____ Advanced: pp. 186–188 Exs. 20, 21, 26, 27, 32, 34, 37–43, 46–53, Quiz 3; EC: CRB p. 117

Reteaching the Lesson

_____ Practice Masters: CRB pages 111–112 (Level A, Level B)

_____ Reteaching with Practice: CRB pages 113–114 or Practice Workbook with Examples; Resources in Spanish

_____ Personal Student Tutor: CD-ROM

Extending the Lesson

_____ Interdisciplinary/Real-Life Applications: CRB page 116

_____ Challenge: CRB page 117

ASSESSMENT OPTIONS

_____ Daily Quiz (3.9): TE page 188 or Transparencies

_____ Standardized Test Practice: SE page 188; STP Workbook; Transparencies

_____ Quiz 3.7–3.9: SE page 188

Notes _____

_____

_____

Copyright © McDougal Littell Inc.
All rights reserved.

Algebra 1
Chapter 3 Resource Book

107

Lesson 3.9

TEACHER'S NAME _____ CLASS _____ ROOM _____ DATE _____

Lesson Plan for Block Scheduling

Half-block lesson (See *Pacing the Chapter,* TE pages 572C–572D)   For use with pages 590–596

GOAL   Solve percent problems.

State/Local Objectives _____

_____

_____

✓ **Check the items you wish to use for this lesson.**

STARTING OPTIONS
_____ Homework Check (3.8): TE page 180; Answer Transparencies
_____ Homework Quiz (3.8): TE page 182, CRB page 109, or
       Transparencies
_____ Warm-Up: CRB page 109 or Transparencies

TEACHING OPTIONS
_____ Lesson Opener: CRB page 110 or Transparencies
_____ Examples 1–4: SE pages 183–185
_____ Extra Examples: TE pages 184–185 or Transparencies; Internet Help at *www.mcdougallittell.com*
_____ Checkpoint Exercises: SE pages 184–185
_____ Concept Check: TE page 185
_____ Guided Practice Exercises: SE page 186

APPLY/HOMEWORK

Homework Assignment (See also the assignment for Lesson 3.8.)
_____ Block Schedule: pp. 186–188 Exs. 18, 19, 24, 25, 30, 31, 34–36, 40–47, 52, Quiz 3

Reteaching the Lesson
_____ Practice Masters: CRB pages 111–112 (Level A, Level B)
_____ Reteaching with Practice: CRB pages 113–114 or Practice Workbook with Examples;
       Resources in Spanish
_____ Personal Student Tutor: CD-ROM

Extending the Lesson
_____ Interdisciplinary/Real-Life Applications: CRB page 116
_____ Challenge: CRB page 117

ASSESSMENT OPTIONS
_____ Daily Quiz (3.9): TE page 188 or Transparencies
_____ Standardized Test Practice: SE page 188; STP Workbook; Transparencies
_____ Quiz 3.7–3.9: SE page 188

Notes _____

_____

_____

CHAPTER PACING GUIDE	
Day	**Lesson**
1	3.1 (all)
2	3.2 (all); 3.3 (all)
3	3.4 (all)
4	3.5 (all); 3.6 (begin)
5	3.6 (end); 3.7 (all)
6	3.8 (all); **3.9 (all)**
7	Ch. 3 Review and Assess

NAME _____ DATE _____

WARM-UP EXERCISES

For use before Lesson 3.9, pages 183–188

Write each number as a percent. Round to the nearest tenth of a percent if necessary.

1. 0.24   **2.** 3.16   **3.** $\dfrac{4}{5}$ **4.** $\dfrac{1}{3}$

··

DAILY HOMEWORK QUIZ

For use after Lesson 3.8, pages 177–182

1. Write the ratio in simplest form: $\dfrac{48}{64}$

2. A football team won 6 out of 10 games. Find the ratio of wins to losses.

3. Find the unit rate: $3 for 12 cans of soda

4. Convert 6 feet to inches.

5. A cheetah can run at a rate of 70 miles per hour. Use unit analysis to find the cheetah's speed in miles per minute. Round your answer to the nearest tenth.

NAME _____ DATE _____

Calculator Lesson Opener

For use with pages 183–188

1. Complete the "Percent as a decimal" column of the table.

	Percent	*Percent as a decimal*	*Number n*	*Percent · n*
Sample	15%	0.15	20	$(0.15)(20) = 3$
Row 1	25%		12	
Row 2	50%		24	
Row 3	75%		16	
Row 4	80%		20	

2. Use a calculator to complete the last column of your table with an equation, as was done in the Sample row.

3. Suppose you do not know the number 12 in Row 1. Replace 12 with b and write an equation. Describe how you could solve this equation.

4. Suppose you do not know the percent in Row 2. Replace 50% with p and write an equation. Describe how you could solve this equation.

5. Suppose you did not know the value of 75% of 16 that you calculated in the last column of Row 3. Replace 12 with a and write a new equation. Describe how you could solve this equation.

Algebra 1
Chapter 3 Resource Book

Lesson 3.9

NAME _____ DATE _____

Practice A

For use with pages 183–188

Write each percent as a decimal.

1. 28%   **2.** 50%   **3.** 80%   **4.** 15%

5. 4%   **6.** 8.5%   **7.** 122%   **8.** 0.5%

Solve the percent problem.

9. What number is 15% of 60?

10. 24 is what percent of 200?

11. 66 is 11% of what number?

12. What number is 32% of 500?

13. 6 is 5% of what number?

14. 16 is what percent of 25?

15. 308 is what percent of 350?

16. 36 is 45% of what number?

17. What number is 60% of 300?

18. 18 is 25% of what number?

19. 51 is what percent of 85?

20. What number is 150% of 80?

In Exercises 21–24, what percent of the region is shaded?

21.
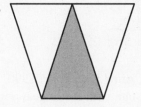

22.
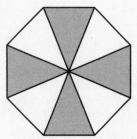

23.
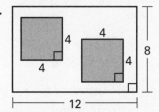

24.
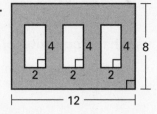

In Exercises 25–28, use the information at the right.

Two hundred people were asked how many hours
a week they exercise.

25. How many people exercised 6–7 hours per week?

26. How many people exercised 4–5 hours per week?

27. How many people exercised 2–3 hours per week?

28. How many people exercised 0–1 hour per week?

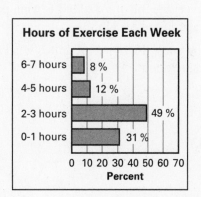

Algebra 1
Chapter 3 Resource Book

111

NAME _____ DATE _____

Practice B

For use with pages 183–188

Solve the percent problem.

1. What number is 73% of 300?

2. What number is 66% of 210?

3. 110 is what percent of 250?

4. 42 is 60% of what number?

5. 36 is 15% of what number?

6. 18 is what percent of 30?

7. 384 is what percent of 480?

8. 121 is 55% of what number?

9. What number is 125% of 124?

10. 52 is 8% of what number?

11. 128 is what percent of 80?

12. What number is 5% of 48?

What percent of the region is shaded? What percent is not shaded?

13.

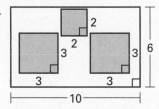

14.

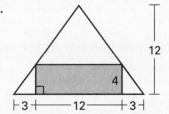

15. ***Earth's Surface Area*** Earth is approximately 30% land. There are approximately 57,280,000 square miles of land on Earth. What is the total surface area of Earth?

16. ***Population of the United States*** In 1990, California was the state with the largest population, 2.984×10^7 people. This was approximately 12% of the total population of the United States. Estimate the population of the United States in 1990.

In Exercises 17–20, use the information at the right.

The table below gives the average life of a Federal Reserve Note.

17. The average life of a $1 bill is what percent of the average life of a $5 bill?

18. The average life of a $1 bill is what percent of the average life of a $10 bill?

19. The average life of a $20 bill is what percent of the average life of a $5 bill?

20. The average life of a $100 bill is what percent of the average life of a $1 bill?

Denomination	*Average Life*
$1	18 months
$5	2 years
$10	3 years
$20	4 years
$50	9 years
$100	9 years

NAME _____ DATE _____

Reteaching with Practice

For use with pages 183-188

GOAL Solve percent problems.

VOCABULARY

A **percent** is a ratio that compares a number to 100. In any percent equation the **base number** is the number that you are comparing to.

EXAMPLE 1 *Number Compared to Base is Unknown*

What is 40% of 65 meters?

SOLUTION

Verbal Model $\boxed{a}$ is $\boxed{p \text{ percent}}$ of $\boxed{b}$

Labels   Number compared to base $= a$ (meters)

                    Percent $= 40\% = 0.40$ (no units)

                    Base number $= 65$ (meters)

Algebraic Model $a = (0.40)(65)$

                     $a = 26$ 26 meters is 40% of 65 meters.

Exercises for Example 1

1. What is 24% of $30?   **2.** What is 60% of 15 miles?

EXAMPLE 2 *Base Number is Unknown*

Twenty-five miles is 20% of what distance?

SOLUTION

Verbal Model $\boxed{a}$ is $\boxed{p \text{ percent}}$ of $\boxed{b}$

Labels   Number compared to base $= 25$ (miles)

                    Percent $= 20\% = 0.20$ (no units)

                    Base number $= b$ (miles)

Algebraic Model $25 = (0.20)b$

                     $\dfrac{25}{0.20} = 125 = b$ 25 miles is 20% of 125 miles.

Exercises for Example 2

3. Sixty grams is 40% of what weight?   **4.** Fifteen yards is 30% of what distance?

Lesson 3.9

Reteaching with Practice

For use with pages 183-188

EXAMPLE 3 *Percent is Unknown*

Ninety is what percent of 15?

SOLUTION

Verbal Model $\boxed{a}$ is $\boxed{p \text{ percent}}$ of $\boxed{b}$

Labels   Number compared to base $= 90$ (no units)

Percent $= p\% = \dfrac{p}{100}$ (no units)

Base number $= 15$ (no units)

Algebraic Model $90 = \dfrac{p}{100}(15)$

$\dfrac{90}{15} = \dfrac{p}{100}$

$600 = p$

90 is 600% of 15.

Exercises for Example 3

5. Forty-five is what percent of 180?   **6.** Sixty is what percent of 15?

EXAMPLE 4 *Modeling and Using Percents*

You took a multiple-choice exam with 200 questions. You answered 80% of the questions correctly. How many questions did you answer correctly?

SOLUTION

You can solve the problem by using a proportion. Let n represent the number of correct answers.

$\dfrac{\textit{Number of correct answers}}{\textit{Total number of answers}} = \dfrac{80}{100}$ Write proportion.

$\dfrac{n}{200} = \dfrac{80}{100}$ Substitute.

$100n = 200 \cdot 80$ Use cross products.

$n = \dfrac{200 \cdot 80}{100}$ Divide by 100.

$n = 160$ Simplify.

You answered 160 questions correctly.

Exercise for Example 4

7. Rework Example 4 if you answered 85% of the questions correctly.

Quick Catch-Up for Absent Students

For use with pages 183–188

The items checked below were covered in class on (date missed) _____

Lesson 3.9: Percents

____ **Goal:** Solve percent problems.

Material Covered:

    ____ Example 1: Number Compared to Base is Unknown

    ____ Student Help: Study Tip

    ____ Example 2: Base Number is Unknown

    ____ Example 3: Percent is Unknown

    ____ Example 4: Model and Use Percents

Vocabulary:

    percent, p. 183   base number, p. 183

____ Other (specify)_____

_____

Homework and Additional Learning Support

    ____ Textbook (specify) <u>pp. 186–188</u>_____

    _____

    ____ Internet: Extra examples at www.mcdougallittell.com

    ____ *Reteaching with Practice* worksheet (specify exercises) _____

    ____ *Personal Student Tutor* for Lesson 3.9

NAME _____ DATE _____

Interdisciplinary Application

For use with pages 183–188

Markup and Cost

ECONOMICS The price of a good or service must be competitive and attractive to consumers, while also covering expenses and providing a reasonable profit for a business.

Expenses are broken down into two major categories: the cost of goods sold and operating expenses or overhead. Operating expenses and profit are combined and referred to as markup. Markup M is the amount of dollars added to the cost C of the goods. Together they equal the selling price S of the item and can be represented by the equation $S = C + M$.

Companies will often determine the amount of markup dollars using a desired markup percent or rate. There are two ways to calculate the markup rate: one is based on cost and the second is based on selling price.

A company may have a set price for the goods they are buying. If this is the case, they have a fixed cost and may determine the markup rate based on cost. For example, a store purchases cereal for $1.50 a box and desires a markup rate R of 80% based on cost. Using the formula $M = RC$ the markup and then the selling price can be determined.

$M = RC$ *Markup based on cost.* $\qquad$ $S = C + M$

$M = (0.80)(1.50) = 1.20$ $\qquad$ $S = 1.50 + 1.20 = 2.70$

On the other hand, a company may have a predetermined selling price. For example, a hamburger franchise that must sell hamburgers at $1.29 has a set selling price. If it desires a markup rate of 45% based on selling price, they could then determine the amount of money they could spend to purchase supplies or its cost.

$M = RS$ *Markup based on selling price.* $\qquad$ $S = C + M$

$M = (0.45)(1.29) = 0.58$ $\qquad$ $1.29 = C + 0.58$

$\qquad\qquad\qquad\qquad\qquad\qquad\qquad\qquad$ $0.71 = C$

Although the calculations are identical, there are actually two different markup rates and therefore two different markups.

1. Your school store is selling pencils. The store paid $.15 per pencil and needs a markup of $.05 each. Determine the selling price.

2. Sweatshirts with your school mascot cost the school $20.00 each. If the store marks up the shirts 35% based on cost, determine the markup and selling price.

3. Backpacks are sold at the store for $35. If the store paid $25 each, determine the markup and the percent of markup based on cost.

4. In Exercise 3, if the markup had been based on the selling price, what would the percent markup be? Round to the nearest tenth.

5. Baseball hats are sold for $15 each. If the store marks them up $7, determine the cost and the percent of the markup based on cost.

NAME _____ DATE _____

Challenge: Skills and Applications

For use with pages 183–188

In Exercises 1–4, solve the percent problem.

1. $\frac{1}{2}$ is what percent of $\frac{3}{4}$?

2. How much is 30% of $\frac{2}{3}$?

3. $\frac{3}{8}$ is 19% of what number?

4. $\frac{4}{5}$ is what percent of $\frac{10}{7}$?

In Exercises 5–7, use the following information.

The length and width of a map that measures 14 inches by 20 inches are each decreased by a scale factor of $\frac{3}{4}$. The scale factor of $\frac{3}{4}$ tells you that each new dimension is $\frac{3}{4}$ of the original dimension.

5. What is the area of the reduced image of the map?

6. By what scale factor is the area of the map changed? Give your answer as a fraction in lowest terms.

7. Based on the answer from Exercise 6, state a general rule about how the area of an image is changed when its length and width are changed by scale factor p.

In Exercises 8–9, use the following information.

In her work for a magazine publisher, Leona Saunders scanned a photograph into electronic form and then loaded it into a graphics editing program, where she reduced the size of the photograph by 40% so that it would fit on a page of the magazine. This size turned out to be too small, so she increased the size of the reduced image by 20%.

8. Find the scale factors of the two size changes that Leona performed. Then calculate the scale factor of the final size in relation to the original size. Give answers in decimal form.

9. A photograph of an image is reduced by x% and then increased by y%. Write an expression in terms of x and y that gives the scale factor of the net size change.

10. The size of an image is increased by 25%. What further percent increase in size will make the net scale factor of the final image $\frac{3}{2}$ in relation to its original size?

NAME _____ DATE _____

Chapter Review Games and Activities

For use after Chapter 3

Solve the following problems in the space provided, and find the answer in the boxes at the bottom of the page. Cross out the box containing each correct answer. Place the remaining letters on the lines below the boxes to find the answer to the riddle.

What do you call emergency relief to Antarctica?

1. $-4 = x - (-5)$

2. $20 = -\dfrac{2}{3}x$

3. $7(x - 3) = 56$

4. $5x - 16 = 3x$

5. $15 - 9 = -x$

6. $12 + \dfrac{x}{5} = -7$

7. $17 = 33 - 8x$

8. $-6(8 - x) + 15 = -51$

9. $-4 + x = 15$

10. $\dfrac{7}{8}x - 22 = 27$

11. $\dfrac{x}{3} = 8$

12. $\dfrac{x}{6} = \dfrac{16}{3}$

13. $6x + 5(x - 2) = 34$

14. $\dfrac{7}{8}(32 + 16x) = -7x - 14$

15. $4x - 17 = 11$

16. $\dfrac{3}{2}x + 5 = 20$

17. $\dfrac{3}{8}x + 21 = 0$

$x = -30$	$x = 4$	$x = -2$	$x = 30$	$x = 19$	$x = -56$
G	I	N	C	A	F
$x = 32$	$x = -11$	$x = 2$	$x = 56$	$x = 6$	$x = -95$
E	O	B	R	O	C
$x = -3$	$x = -9$	$x = 3$	$x = -24$	$x = 11$	$x = 7$
U	D	L	A	N	H
$x = -32$	$x = 24$	$x = 8$	$x = 10$	$x = -6$	$x = 9$
I	G	P	T	E	D

___ ___ ___ ___   ___ ___ ___!!

Algebra 1
Chapter 3 Resource Book

Chapter Test A

For use after Chapter 3

Solve the equation.

Answers

1. $x - 12 = 13$ **2.** $-16 = 9 + x$

3. The selling price of a pair of pants is $34. If the store paid $8 less for the pants, find the price the store paid.

Tell whether the equations are equivalent.

4. $17x = 85$ and $x = 5$ **5.** $-16x = 48$ and $x = 3$

Solve the equation.

6. $35a = -70$ **7.** $\dfrac{b}{2} = 14$

8. You estimate that you spend $115 on groceries each month. How much money do you spend on groceries each week?

Solve the equation.

9. $\frac{1}{4}x - 5 = 27$ **10.** $14x + 3x - 40 = 11$

11. *Consecutive integers* are integers that follow each other in order (for example, 5, 6, and 7). The sum of three consecutive integers is 417. Write and solve an equation that will determine the three integers. Let x be the first integer.

Solve the equation.

12. $3x - 8 = -3x + 4$ **13.** $40 - 14y = 6y$

14. A local gym charges nonmembers $10 per hour to use the tennis courts. Members pay a yearly fee of $300 and $4 per hour for using the tennis courts. Write and solve an equation to find how many hours you must use the tennis courts to justify becoming a member.

15. You are designing a web page for your school's biology club. You want to include photos of the members on the page, which has a width of 640 pixels. You've decided that the left and right margins should be 24 pixels each and the space between each picture should be 16 pixels. How wide can each picture be to fit four across the width of the page?

Perform any indicated operation. Round the result to the nearest tenth and then to the nearest hundredth.

16. $-49.256 + 17.197$ **17.** $14.357(-2.625)$

Answers

1. _____
2. _____
3. _____
4. _____
5. _____
6. _____
7. _____
8. _____
9. _____
10. _____
11. _____
12. _____
13. _____
14. _____
15. _____
16. _____
17. _____

Solve the equation. Round the result to the nearest hundredth.

18. $-15x + 73 = 26$ **19.** $7y - 27 = 14$

20. $1.5x + 7.4 = -1.8x - 6.9$ **21.** $14.1 - 0.3x = 1.3x - 4.2$

22. $6.2x + 4.5 = 3.8x + 7.9$ **23.** $4.603y - 1.842 = -3.651y$

Solve for the indicated variable.

24. Area of a Trapezoid

Solve for b_1: $A = \frac{1}{2}h(b_1 + b_2)$

25. Simple Interest Formula

Solve for t: $I = Prt$

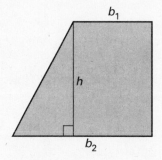

Solve the equation for *y*.

26. $7x + y = 13$ **27.** $3x + 5y = 15$

28. Use the result in Exercise 26 to find y when $x = -1, 0,$ and 2.

29. Use the result in Exercise 27 to find y when $x = -1, 0,$ and 2.

Find the unit rate.

30. 3 tablespoons for 1.5 servings

31. $10.99 for 12 slices of pizza

Convert the measure. Round your answer to the nearest tenth.

32. 15 teaspoons to tablespoons (1 tablespoon = 3 teaspoons)

Find the percent. Round to the nearest whole percent.

33. 159 people in favor out of 350 people surveyed

| 18. _____ |
| 19. _____ |
| 20. _____ |
| 21. _____ |
| 22. _____ |
| 23. _____ |
| 24. _____ |
| 25. _____ |
| 26. _____ |
| 27. _____ |
| 28. _____ |
| 29. _____ |
| 30. _____ |
| 31. _____ |
| 32. _____ |
| 33. _____ |

Review and Assess

NAME _____ DATE _____

Chapter Test B

For use after Chapter 3

Solve the equation.

1. $14 = x - (-7)$

2. $3 - (-x) = 19$

3. You owed $34 to your sister. You paid x dollars back and you now owe $12. How much did you pay back?

Tell whether the equations are equivalent.

4. $\frac{4}{5}x = 50$ and $x = 32$

5. $\frac{2}{3}y = -48$ and $y = -72$

Solve the equation.

6. $\frac{3}{8}y = 12$

7. $-\frac{5}{7}t = -25$

8. $\frac{4}{9}a = 8$

Solve the equation.

9. $5x + 7 - 2x = 22$

10. $12x - (4x + 10) = 54$

11. $8x - 10(3 - x) = 42$

12. $\frac{2}{3}(x + 6) = 8$

13. The sum of three numbers is 137. The second number is 4 more than two times the first number. The third number is 5 less than three times the first number. Find the three numbers.

Solve the equation if possible.

14. $4(x - 5) = 4x - 20$

15. $6(x - 9) = -12x + 36$

16. A local gym charges nonmembers $8 per day to use the volleyball courts. Members pay a yearly fee of $150 and $2 per day to use the volleyball courts. Write and solve an equation to find how many days you must use the volleyball courts to justify becoming a member.

Answers

1. _____
2. _____
3. _____
4. _____
5. _____
6. _____
7. _____
8. _____
9. _____
10. _____
11. _____
12. _____
13. _____
14. _____
15. _____
16. _____

Perform any indicated operation. Round the result to the nearest tenth and then to the nearest hundredth.

17. $-57.124 \div 10.104$ **18.** $-13.254(-3.145)$

Solve the equation. Round the result to the nearest hundredth.

19. $1.59x + 4.23 = 3.56x + 2.12$

20. $-3(1.25 - 2.48x) = 8.15x + 5.86$

21. $4.15x + 3.01 = 10.9x + 1.29$ **22.** $1.596y - 3.08 = 0.9y$

Solve for the indicated variable.

23. Volume of a Cone

Solve for h: $V = \dfrac{\pi r^2 h}{3}$

24. Temperature Formula

Solve for C: $F = \frac{9}{5}C + 32$

Solve the equation for *y*.

25. $\frac{2}{3}y + 4 = 2x$ **26.** $\frac{1}{2}(y + 5) + 4x = 3x$

27. Use the result in Exercise 25 to find y when $x = -1$, 0, and 2.

28. Use the result in Exercise 26 to find y when $x = -1$, 0, and 2.

29. A store sells 28 ounces of peanut butter for $2.24. The store also sells 32 ounces of the same peanut butter for $2.40. Which is the better buy?

In Question 30, convert the measure. Round your answer to the nearest tenth.

30. 147 miles to kilometers (1 mile = 1.609 kilometers)

31. A waiter typically receives about 15% of a total food bill as a tip. To earn a total of $50 in tips, about how much food would the waiter have to sell?

17. _____

18. _____

19. _____

20. _____

21. _____

22. _____

23. _____

24. _____

25. _____

26. _____

27. _____

28. _____

29. _____

30. _____

31. _____

NAME _____ DATE _____

SAT/ACT Chapter Test

For use after Chapter 3

1. Solve $-7 = 4 - (-x)$.

   (A) -11 (B) -3

   (C) 3 (D) 11

2. Which one of these steps can you use to solve the equation $14 = \frac{2}{3}x$?

   I.   Multiply by $\frac{2}{3}$ II.   Divide by $\frac{2}{3}$

   III. Multiply by $\frac{3}{2}$ IV. Divide by $\frac{3}{2}$

   (A) I only   (B) III only

   (C) I and IV   (D) II and III

3. If $\frac{2}{5}x = -\frac{10}{13}$, then $x =$ _____?

   (A) $-\frac{4}{13}$ (B) $-\frac{13}{25}$

   (C) $-\frac{25}{13}$ (D) $-\frac{13}{4}$

4. Solve the equation $\frac{2}{3}x + 5 = 13$.

   (A) 6 (B) 8

   (C) 12 (D) 27

5. If $2x - 4(3 - x) = 18$, then $x =$ _____?

   (A) -15 (B) -3

   (C) 1 (D) 5

6. Find the value of x if $6(2 - x) + 4x = -5(x + 3)$.

   (A) $-\frac{9}{5}$ (B) $-\frac{7}{3}$

   (C) $-\frac{27}{7}$ (D) -9

7. Which equations are equivalent?

   I. $8x - 3 = 12$ II. $-12 = 3 - 6x$

   III. $3(2x - 4) = 3 - 2x$

   (A) I and III   (B) II and III

   (C) All   (D) None

8. The sales tax rate is 0.06. What is the total bill for your meal before tax if the sales tax came to $.51?

   (A) $1.18   (B) $8.50

   (C) $9.01   (D) $11.76

In Questions 9 and 10, choose the statement below that is true about the given numbers.

   a. The number in column A is greater.
   b. The number in column B is greater.
   c. The two numbers are equal.
   d. The relationship cannot be determined from the given information.

9.

Column A	Column B
x when $7x - 5 = 12$	3

   (A)   (B)   (C)   (D)

10.

Column A	Column B
16	y when $\frac{1}{2}y - 3 = 5$

   (A)   (B)   (C)   (D)

11. Solve $1.69x + 14.75 = 4.21x - 5.87$.

   (A) 0.12 (B) 0.28

   (C) 3.52 (D) 8.18

12. Use the equation $3(4x - 2y) = 5$. What are the values of y when $x = -2, 0, \frac{1}{3}$, and 3?

   (A) $-\frac{22}{3}, -\frac{10}{3}, -\frac{8}{3}, \frac{8}{3}$

   (B) $-\frac{7}{12}, \frac{5}{12}, \frac{7}{12}, \frac{23}{12}$

   (C) $-\frac{29}{6}, -\frac{5}{6}, -\frac{1}{6}, \frac{31}{6}$

   (D) $-\frac{58}{3}, -\frac{10}{3}, -\frac{2}{3}, \frac{62}{3}$

Review and Assess

JOURNAL **1.** A classmate has just solved an equation without checking the answer. Unfortunately, the solution has several mistakes. Without re-doing the entire problem, go through each step and explain how the step was obtained. If the step is incorrect, explain why it's an error and explain the correct procedure that should have been followed.

$$7 + \tfrac{1}{2}(6x + 4) = -2x + 4 \qquad \text{Original equation}$$

$$7 + 3x + 2 = -2x + 4$$

$$10x + 2 = -2x + 4$$

$$8x + 2 = 4$$

$$8x = 2$$

$$x = 4$$

MULTI-STEP PROBLEM **2.** Ron the realtor is offered a job directly out of real-estate school. He has a choice as to which way he will receive his salary the first year.

Salary Plan 1: He would receive a base pay of $2000 per month plus a 3% commission on each sale.

Salary Plan 2: No base pay but a 6% commission on each sale.

a. If Ron averages $45,000 in sales per month, how much would he earn under the first plan? How much under the second plan? Which is the better choice in this case?

b. Write a verbal model for Salary Plan 1 that relates base pay, total pay, sales in a month, and percent commission.

c. Write a verbal model for Salary Plan 2 relating sales in a month, percent commission, and total pay.

d. Write an equation to determine when it would be better to switch from the first plan to the second plan. Give a one- or two-sentence answer that includes Ron's sales in a month. Round to the nearest dollar, if necessary.

e. Ron makes a goal to earn $8000 every month. State which choice he should select and explain your reasoning. Use the verbal model from (b) or (c) to set up an equation and determine what his sales should be to achieve this goal.

3. *Critical Thinking* Suppose Ron's salary goal for the year is $50,000. If the average price of a house sold by his company is $60,000, how many houses would Ron have to sell over the course of the year to meet his goal under each plan?

Alternative Assessment and Math Journal

For use after Chapter 3

JOURNAL SOLUTION

1. Complete answers should include these points:

$7 + \frac{1}{2}(6x + 4) = -2x + 4$	Original equation
$7 + 3x + 2 = -2x + 4$	This step is correct.
$10x + 2 = -2x + 4$	Error: added 7 and $3x$ to get $10x$; incorrect because 7 and $3x$ are not like terms. The correct method is to add like terms 7 and 2, to get $3x + 9$.
$8x + 2 = 4$	Error: incorrect to add $10x$ and $-2x$ because they are on different sides of equation. The correct method is to add $2x$ to *both* sides of the equation.
$8x = 2$	This step is correct.
$x = 4$	Error: did not divide both sides by 8 to isolate the variable. The correct method is to divide *both* sides by 8 to isolate the variable.

MULTI-STEP PROBLEM SOLUTION

2. a. $3350; $2700; Salary Plan 1

b. | total pay | $=$ | base pay | $+$ | percent commission | $\cdot$ | sales in a month |

c. | total pay | $=$ | percent commission | $\cdot$ | sales in a month |

d. $2000 + 0.03x = 0.06x$; Answers should include these points:
If Ron's sales per month are $66,666 or less, he should choose Plan 1.
If his sales per month are greater than $66,667, he should choose Plan 2.

e. Answers should include the following points and equation:
Ron should choose Plan 2 because monthly sales must exceed $66,667 in order to provide a salary of $8,000 a month. Since $8000 = 0.06x$ (where $x =$ sales per month), $x = 133,333$. Ron must make $133,333 in sales each month.

3. 15 houses under Plan 1; 14 houses under Plan 2

MULTI-STEP PROBLEM RUBRIC

4 Students accurately compute monthly salaries and write correct verbal models. They make a reasonable hypothesis based on their possible monthly salaries. Students use correct equations to answer questions, and then compute the salary for when the choices are equivalent. Students demonstrate an understanding of when to use which choice and give clear explanations.

3 Students are able to correctly set up models and equations; however, there are errors in the solution. Their hypothesis is reasonable even if based on incorrect salaries. Explanation of reasoning is incomplete or not all aspects of each question are answered.

2 Students have an incorrect model for computing salaries or set up incorrect equation(s). Students do not give a clear explanation as to when to use which choice, and do not answer all parts of each question.

1 Students' work is very incomplete. They are unable to create models or set up equations and show little understanding of how to compute salaries. Minimal explanation is provided.

Review and Assess

Project: Ice Rescue

For use after Chapter 3

OBJECTIVE Determine the safest and best ways to rescue a child stranded on thin ice

MATERIALS paper, pencil, calculator, quarter-inch grid paper, sheet of ice at least 1-inch thick formed in a serving dish or baking pan, empty soup can, heavy objects like books, scale

INVESTIGATION A 100-pound child is stranded in the middle of a pond on thin ice. You have three options to use to rescue the child. In all three, assume you carry the child back.

- You can slide on your feet.

- You can use your ice skates which each weigh one pound more than a shoe. The area of a blade that touches the ice is 2 square inches.

- You can lay down on a 9-foot-by-2-foot board and slide. The board weighs 10 pounds.

Stress (σ) can be found by the formula $\sigma = \dfrac{F}{A}$, where F is the force and A is the area on which the force acts. In this situation, force is equal to the weight being put on the ice and the area is the area of contact with the ice.

1. Determine how many pounds per square inch (psi) of stress your sheet of ice can handle. You will use this number to represent the stress of the ice in the pond. Prop the ice so the center of the sheet is not supported. Put the can in the center and then stack books on the can, one at a time, until the ice breaks. Weigh the can together with all but the last book. Substitute this value into F in the stress formula. Find the area of the can that touched the ice and then find the stress. (*Hint:* The area of a circle is πr^2, where r is the radius of the circle.)

2. Use the quarter-inch grid paper to estimate the area of your foot. Write an equation for the maximum you can weigh to safely rescue the child by sliding on your feet. Solve the equation.

3. Write an equation for the maximum you can weigh to safely rescue the child using ice skates. Assume you keep both skates on the ice at all times. Solve the equation.

4. Write an equation for the maximum you can weigh to safely rescue the child by sliding on the board. Solve the equation.

Make a poster to present your results. Include your equations. Make diagrams illustrating the stress involved in each option. Decide which option is the safest and explain. Decide if the safest option is the best for rescuing the child and explain.

Project: Teacher's Notes

For use after Chapter 3

GOALS • Use two or more transformations to solve an equation and use multi-step equations to solve real-life problems.

• Use rates and ratios to model and solve real-life problems.

• Collect and analyze data.

MANAGING THE PROJECT The sheet of ice can be formed by freezing water in a shallow serving dish or baking pan. The sheet of ice needs to be long enough to reach between whatever supports are used and wider than the can.

To estimate the area of the bottom of their feet, have students trace one of their feet onto the quarter-inch grid paper, count the whole squares, and combine partial squares. You may want to remind them to divide the total number of squares by 16 to find the area in square inches.

Be sure students include the total weight in each equation. They may omit the weight of the child, the ice skates, or the board.

RUBRIC The following rubric can be used to assess student work.

4 The student collects the data carefully and analyzes it correctly, uses equations to determine the maximum weight for each rescue option, and decides the safest and the best options. The poster presents the best options, makes a convincing case for decisions, and shows supporting evidence.

3 The student collects and analyzes data, uses equations to determine the maximum weight for each rescue option, and decides the safest and the best options. However, the student may make errors in collecting data, may not perform all calculations correctly, or may not fully address all issues when choosing the safest and best options. The poster gives and supports opinions about the best options, but the presentation may not be as convincing as possible.

2 The student collects data, solves equations, and presents an opinion about the safest and best options. However, work may be incomplete or reflect misunderstanding. For example, the student may use inappropriate procedures to collect the data or may make errors in setting up and solving the equations. The poster may indicate a limited grasp of certain ideas or may lack key supporting evidence.

1 Data analysis, equations, and decisions on the best options are missing or do not show an understanding of key ideas. The poster does not give a reasonable decision or fails to support the decision.

Review and Assess

Cumulative Review

For use with Chapters 1–3

Evaluate the expression for the given value(s) of the variable(s). (1.2)

1. $(x - y)^2$ when $x = 3$ and $y = -4$

2. $a + b^3$ when $a = 3.4$ and $b = 6$

3. $2(d - 5)^2$ when $d = 8$

4. $10 - (-x)^3 + y^2$ when $x = -1$ and $y = -4$

Check whether the given number is a solution of the inequality. (1.4)

5. $c - 7 < 9;\ 15$

6. $9 + 9x \geq 18;\ 1$

7. $14 \leq -(2x - 7);\ 20$

8. $y^3 - 9 \leq 27;\ 3$

9. $n(34 - n) < 111;\ -3$

10. $\dfrac{c + 7}{9} \leq 7;\ 100$

Write the verbal phrase as an equation or an inequality. Use *x* for the variable in your expression. (1.5)

11. Four is greater than five times a number.

12. Twelve less than the product of eight and a number is less than fifty.

13. The product of one half and a number is greater than or equal to one hundred.

14. Two and five ninths decreased by a number is thirty-one.

Based on the graph, decide whether each statement is *true* or *false*. (1.7)

15. Troy's bowling score increased each week.

16. Troy's score was the same week 3 and week 5.

17. Troy's high schore was 275.

18. Troy's score exceeded his average weeks 3, 5 and 6.

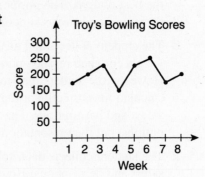

Write the numbers in increasing order. (2.1)

19. $3, -6, 0, \frac{12}{8}, 2, -\frac{10}{3}$

20. $8.6, 8.599, 8.023, 8.06, 8.5, -8.9$

21. $3\frac{2}{5}, 3.25, -3, -3.023, -3.6, -3\frac{6}{7}$

22. $9, -\frac{4}{17}, -\frac{34}{9}, -8.9, -9.01, -1$

Find the sum or the difference. (2.2, 2.3)

23. $7.90 + (-12.01) + 9.14$

24. $\frac{2}{3} + 9 - \frac{1}{24}$

25. $7.23 - (-2.34) - 0.001$

26. $\frac{7}{15} - \frac{1}{120} + \frac{3}{4}$

27. $-10 - (-0.99)$

28. $(3 - 2.56) - (5.7 + 1.1)$

Solve the equation. (3.1)

29. $-8 = 3 - y$

30. $\frac{4}{9} = s - \frac{1}{18}$

31. $r + 6\frac{1}{5} = 10\frac{2}{15}$

32. $y - (-8.3) = 7.7$

33. $11 + x = 11$

34. $5.1 - (-2.1) = -y$

Cumulative Review

For use with Chapters 1–3

Solve the equation. (3.2)

35. $\dfrac{y}{3} = -17$

36. $0 = \dfrac{18}{7}t$

37. $42b = 7$

38. $\dfrac{1}{9}y = 12$

39. $-\dfrac{6}{7}x = 2\dfrac{1}{8}$

40. $-5a = -30$

Check whether the given number is a solution of the equation. (3.3)

41. $2x - x - 23 = -2, 7$

42. $\dfrac{5}{6}x + 2 = -8, 12$

43. $7x - 6(3 - x) = 26, 8$

44. $\dfrac{x}{3} - 4 = 5, 27$

Solve the equation if possible. (3.4, 3.5)

45. $8 - (-3n) = 3n - 2$

46. $3.8y - 4.7 = 3.8y + 17.5$

47. $9.1(1 - x) + 5x = -4.2(x - 8)$

48. $-2(a + 5) = 27 - 2a$

49. $\dfrac{5}{6}(24 - 36b) = 10(2b + 4)$

50. $-9(x - 3) = -(2 - 9x)$

Solve (3.6)

51. You are in a restaurant and your bill comes to $25. You want to leave a 15% tip. What is your total bill?

52. Five people want to share equally in the cost of a birthday present. The present costs $105.99. How much does each person pay?

Solve the equation. Round your answer to the nearest tenth. (3.6)

53. $3.6x + 0.9 = 8.8 - 1.5x$

54. $-0.389y - 0.974 = 6.789y$

55. $6.43 + 6.58x = 8.68(x - 4)$

56. $0.2n - 0.594 = 3.5n + 9.0$

Solve the equation for y. (3.7)

57. $5x - 2y = 8$

58. $x = 2y + 9$

59. $-2x + 3y = 7$

60. $14x - 7y = 28$

In Exercises 61–64, convert the measure. Round your answer to the nearest tenth. (3.8)

61. 224 days to weeks (1 week $=$ 7 days)

62. 96 inches to centimeters (1 inch $=$ 2.54 centimeters)

63. 1.5 miles to kilometers (1 mile $=$ 1.609 kilometers)

64. $32.50 Canadian dollars to U.S. dollars
(1 U.S. dollar $=$ 1.1515 Canadian dollars)

NAME _____ DATE _____

Cumulative Test 1-3

For use after Chapter 3

In Exercises 1–6, evaluate the expression for the given value of the variable(s). (1.1, 1.2, 2.5, 2.8)

1. $7 + p$ when $p = 9$

2. $(x + y)^2$ when $x = 2$ and $y = 5$

3. $(4m)^3$ when $m = 2$

4. $-5(d)$ when $d = -3$

5. $6(-3)(-s)^2$ when $s = 4$

6. $\dfrac{5x - 4}{-7}$ when $x = -2$

In Exercises 7–13, evaluate the expression. (1.3, 2.3, 2.5, 2.8)

7. $2^3 + 8(7)$

8. $\dfrac{20 + 8}{3^2 + (11 - 6)}$

9. $15 + (-6) - 7$

10. $3(-4)(-2)$

11. $\dfrac{9}{-\frac{1}{3}}$

12. $\dfrac{5}{3} - \dfrac{10}{3}$

13. $(-2)(5 - 7) - 3^2 + 4$

Answers

1. _____

2. _____

3. _____

4. _____

5. _____

6. _____

7. _____

8. _____

9. _____

10. _____

11. _____

12. _____

13. _____

NAME _____ DATE _____

Cumulative Test 1-3

For use after Chapter 3

In Exercises 14–17, write the sentence as en equation or inequality. Let *x* represent the number. (1.5)

14. The sum of a number and 7 is 23.

15. 12 less than a number is 16.

16. 35 is more than the product of 5 and a number.

17. The quotient of a number and 6 is less than 27.

In Exercises 18 and 19, make an input-output table for the function. Use 0, 1, 2, and 3 as values for *x*. Then draw a line graph to represent the function given by the table. (1.8)

18. $y = 3x - 5$

Input	0	1	2	3
Output				

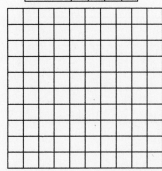

19. $y = 4(2 - x)$

Input	0	1	2	3
Output				

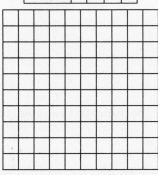

In Exercises 20–22, write the numbers in increasing order. (2.1)

20. $-3.9, 2, -1.4, 0.6, -1.72$

21. $\dfrac{3}{2}, -2, -1.7, -\dfrac{4}{3}, -0.5$

22. $|-6|, -\dfrac{10}{2}, 1, -5.3, |2.7|$

In Exercises 23–27, simplify the expression. (2.7)

23. $-15m + 7m$

24. $2c + 5 - c$

25. $p^2 + 4p + 2p^2 - 6$

26. $3(4r + 1) + 2(r - 7)$

27. $5(x - 9) - (x + 1)$

Answers

14. _____

15. _____

16. _____

17. _____

18. *Use table and grid.*

19. *Use table and grid.*

20. _____

21. _____

22. _____

23. _____

24. _____

25. _____

26. _____

27. _____

Review and Assess

In Exercises 28–30, use the distributive property to rewrite the expression without parentheses. (2.6)

28. $3(x - 7)$

29. $-6(5 + a)$

30. $-9(2a + 4)$

In Exercises 31–37, solve the equation. (3.1–3.5)

31. $\dfrac{1}{3} + x = -\dfrac{2}{3}$

32. $4r = 28$

33. $\dfrac{3}{4}k = 9$

34. $12n + 16 = 40$

35. $x - 3(2x + 1) = 17$

36. $2(x + 1) = 4(x - 3)$

37. $\dfrac{2}{5}(5m + 15) = -16 - 2(m - 3)$

In Exercises 38–40, solve the equation for the indicated variable. (3.7)

38. $x + y = z$ for y

39. $ax - b = c$ for x

40. $4(m + 1) = n$ for m

Answers

28. _____

29. _____

30. _____

31. _____

32. _____

33. _____

34. _____

35. _____

36. _____

37. _____

38. _____

39. _____

40. _____

Review and Assess

In Exercises 41–46, solve the percent problem. (3.9)

41. How much money is 45% of $250?

42. 16% of 180 feet is what length?

43. 20 grams is 60% of what weight?

44. 3% of what amount is $20?

45. 8 people is what percent of 90 people?

46. 10 inches is what percent of 70 inches?

47. You need to buy 5 trees to plant in your back yard. Each one costs $2.30. Use the distributive property to mentally calculate the total cost of the trees. (2.6)

In Exercises 48 and 49, write and solve an equation to solve the problem. (3.1–3.5)

48. There were 25 people at the party. If 17 went home, how many people were left?

49. A new bike costs $360. You have $150 in the bank and earn $30 a week babysitting. How many weeks will it take to make enough money to buy the new bike?

Answers

41. _____

42. _____

43. _____

44. _____

45. _____

46. _____

47. _____

48. _____

49. _____

Review and Assess

ANSWERS

Chapter Support

Parent Guide
Chapter 3

3.1: $x + 37 = 64$; $27

3.2: $\frac{1}{5}x = 136$; 680 students   **3.3:** 2   **3.4:** 3

3.5: $189 + 9.75x = 216 + 5.25x$; 6 CDs or cassettes   **3.6:** -18.78 **3.7:** $x = \frac{7}{3}y - 10$; -17

3.8: about 104.6 km/hr   **3.9:** 360

Prerequisite Skills Review

1. $-5; \frac{1}{5}$ 2. $\frac{2}{3}; -\frac{3}{2}$ 3. $-\frac{1}{5}; 5$ 4. $7; -\frac{1}{7}$

5. $1; -1$ 6. $-10; \frac{1}{10}$ 7. no   8. yes   9. yes

10. no   11. $96 - 8y$ 12. $-6x + 135$

13. $-5.8x^2 + 14.5x$ 14. $3z - \frac{9}{7}y$

15. $5x + 3$ 16. $3m + 28$ 17. $10x - 2y$

18. $20y$ 19. $-5x + 50$ 20. $-14y + 4n$

Strategies for Reading Mathematics

1. The number 95 is the Fahrenheit temperature to be converted. In the formula, F indicates the Fahrenheit temperature. 2. The purpose of Example 6 is to model how to solve problems using a known formula. To find the Fahrenheit temperature, substitute the Celsius temperature given and solve the equation for F, the Fahrenheit temperature.

3. $A = \frac{1}{2}bh$

$20 = \frac{1}{2} \cdot b \cdot 5$

$20 \div \frac{1}{2} = \frac{1}{2} \cdot b \cdot 5 \div \frac{1}{2}$

$20 \cdot 2 = \frac{1}{2} \cdot b \cdot 5 \cdot 2$

$40 = b \cdot 5$

$40 \div 5 = b \cdot 5 \div 5$

$8 = b$

Lesson 3.1

Warm-Up Exercises

1. -5 2. -7 3. -4 4. 10   5. 6

Daily Homework Quiz

1. -10 2. 33   3. $5d - 25$ 4. $-\frac{1}{12}$

5. all real numbers except $x = 3$

Lesson Opener

Allow 15 minutes.

1, 2. Check students' work.

Practice A

1. subtract 12   2. subtract -6 3. add 21

4. add -15 5. subtract -48 6. add -30.2

7. yes   8. no   9. no   10. yes   11. yes

12. yes   13. -4 14. 12   15. 9   16. 21

17. 3   18. 11   19. 3   20. -1 21. 7

22. -11 23. -4 24. 15   25. -7 26. -4

27. 3   28. -10 29. 20   30. -3 31. -6

32. -6 33. 5

34. $x - 3000 = 21,000$; 24,000 feet

35. $x - 5826 = 14,494$; 20,320 feet

36. $x + 3 = 8.3$; 5.3 miles

37. $x - 17 = 28$; 45 days

Practice B

1. subtract 18   2. subtract -11 3. add 31

4. add -25 5. add $-4\frac{1}{2}$ 6. add -10.06

7. -9 8. 5   9. 14   10. -1 11. -3

12. 21   13. -21 14. -24 15. -2

16. $-\frac{5}{4}$ 17. 1   18. $1\frac{1}{2}$ 19. $\frac{1}{3}$

20. 20.1   21. -12.3 22. -18 23. -16

24. -9 25. -24 26. 8   27. 34

28. 13 ft   29. 10 cm

30. $x - 3800 = 28,000$; 31,800 feet

31. $x - 8708 = 20,320$; 29,028 feet

Lesson 3.1 *continued*

32. $x + 2.7 = 8.3$; 5.6 miles

33. $K = C + 273.15$; 257.15°K

Reteaching with Practice

1. 15 **2.** -1 **3.** 13 **4.** -9 **5.** 16 **6.** -6
7. -7 **8.** 9 **9.** 14 **10.** $275

Real-Life Application

1. $93,967 + x = 107,501$; $x = 13,534$

2. $102,854 + x = 107,502$; $x = 4648$

3. $x + 7850 = 93,967$; $x = 86,117$

4. $x - 47,499 = 107,501$; $x = 155,000$

Challenge: Skills and Applications

1. $2, -2$ **2.** $11, -11$ **3.** $5\frac{1}{2}, -5\frac{1}{2}$ **4.** $8, -8$
5. $13, -13$ **6.** $7, -7$ **7.** $6, -6$ **8.** $23, -23$

9.

Length of fourth side		$+ 3 \cdot$	Length of other 3 sides		$=$	Total fencing

$x + 3\left(\frac{3}{8}\right) = 1\frac{3}{8}$ **10.** $\frac{1}{4}$; the fourth side is $\frac{1}{4}$ mi long

11. Taro, Anne, Jamala, Peter

12. $\left(x + 4\frac{1}{4}\right) - \frac{3}{4}$; simplified form: $x + 3\frac{1}{2}$

13. $x + 3\frac{1}{2} = 58$; $54\frac{1}{2}$ in.

Lesson 3.2

Warm-Up Exercises

1. 1 **2.** 1 **3.** 21 **4.** 2 **5.** 1

Daily Homework Quiz

1. -7 **2.** 5 **3.** $\frac{2}{5}$ **4.** 8 **5.** 117 feet

Lesson Opener

Allow 15 minutes.

1. 2 **2.** -3 **3.** 5 **4.** -4

5. To solve a multiplication equation, divide both sides by the coefficient of the variable.

Practice A

1. multiply by 7 **2.** multiply by -4
3. divide by 3 **4.** divide by -6
5. multiply by $-\frac{1}{3}$ **6.** divide by $-\frac{3}{5}$ **7.** $4; \frac{1}{4}$

8. $-\frac{5}{3}; -\frac{3}{5}$ **9.** $2; \frac{1}{2}$ **10.** yes **11.** no **12.** no
13. yes **14.** 3 **15.** -8 **16.** -2 **17.** 6
18. 20 **19.** $\frac{1}{5}$ **20.** -8 **21.** $-\frac{1}{3}$ **22.** $-\frac{1}{2}$
23. 12 **24.** -20 **25.** 30 **26.** $\frac{1}{2}$ **27.** -3
28. $-\frac{2}{3}$ **29.** $4x = 12$; 3 ft **30.** $\frac{1}{4}x = 1\frac{1}{2}$; 6 laps

Practice B

1. multiply by 8 **2.** multiply by -5
3. divide by 10 **4.** divide by $\frac{1}{2}$
5. multiply by $-\frac{2}{3}$ **6.** divide by $-\frac{6}{5}$ **7.** $8; \frac{1}{8}$
8. $-\frac{10}{7}; -\frac{7}{10}$ **9.** $-2; -\frac{1}{2}$ **10.** 13 **11.** -16
12. -6 **13.** 9 **14.** $-\frac{1}{3}$ **15.** $\frac{1}{5}$ **16.** $\frac{2}{3}$
17. 33 **18.** -68 **19.** 120 **20.** 60 **21.** -16
22. $-\frac{5}{2}$ **23.** $-\frac{2}{5}$ **24.** $\frac{26}{3}$ **25.** 0 **26.** 24
27. 4.5 **28.** $\frac{2}{5}x = 130$; 325 students
29. $8x = 78$; $9.75
30. $234x = 2500$, 10 windows;
$11x = 2500$, $227

31. $\frac{5}{8}x = 330$; 528

Reteaching with Practice

1. $-\frac{1}{2}$ **2.** $\frac{1}{6}$ **3.** 7 **4.** -12 **5.** 21 **6.** 18
7. $s(2.5) = 800$; 320 miles per hour
8. $340t = 1530$; 4.5 hours

Interdisciplinary Application

1. $10x = 1966$; 196.6 miles per day

2. $255x = 1966$; about 7.7 days

3. $\frac{x}{9} = 10$; $90 **4.** $\frac{x}{30} = \frac{100}{4}$; $750

Challenge: Skills and Applications

1. $\frac{13}{2}, -\frac{13}{2}$ **2.** $\frac{12}{5}, -\frac{12}{5}$ **3.** $35, -35$
4. $70, -70$ **5.** $\frac{4}{7}$ **6.** $-\frac{1}{10}$ **7.** $-\frac{5}{3}$ **8.** $-\frac{1}{6}$
9. 2.5 **10.** The transformed equation is $0x = 0$, which is true for all x, so it does not have the same solution as $4x = 12$.

11. Division by zero is undefined.

12. *Sample equation:* $\frac{20}{25} + \frac{d}{25} = 2$; 30 mi

Lesson 3.3

Warm-Up Exercises

1. $6 - 18x$ 2. $-5x - 7$ 3. $-14x + 77$
4. $2x + 3$ 5. $-2x + 2$

Daily Homework Quiz

1. -7 2. -9 3. -6 4. 18 5. $\frac{x}{4} = 9.45$

Lesson Opener

Allow 10 minutes.

1. C; The equation shows the charge per service call, 15, added to 25 times the number of hours worked, $25x$, added to the cost for the parts, 65, all set equal to the total bill of 155.

2. A; The equation shows the cost for the software package, 24.99, added to 3 times the cost of one printer cartridge, $3x$, added to the tax, 5.95, all set equal to the total bill of 90.91.

3. D; The equation shows the cost per minute times the number of minutes, $0.09x$, added to the cost for the plan, 12.99, all set equal to the total bill of 21.54.

Graphing Calculator Activity

1. 6 2. -12 3. 10 4. -2 5. 0 6. -24

Practice A

1. no 2. yes 3. yes 4. no 5. yes 6. no
7. Subtract 9 from each side of the equation.
8. Add 5 to each side of the equation.
9. Subtract 6 from each side of the equation.
10. Use the distributive property on $2(3x - 4)$.
11. Combine like terms $3x$ and $8x$.
12. Subtract 8 from each side of the equation.
13. 8 14. 5 15. 4 16. 3 17. 5 18. $-\frac{7}{2}$
19. -14 20. -6 21. 1 22. 1 23. 3
24. 5 25. -3 26. -5 27. -3 28. $-\frac{1}{4}$
29. -6 30. -2
31. $7 \cdot 5 + 1 + x = 88$; 52 white keys
32. $4x + 800 = 2000$; 300 sandwiches

Practice B

1. yes 2. yes 3. yes 4. no 5. no 6. no

---

7. 9 8. 7 9. -2 10. 18 11. 1 12. -24
13. -11 14. -6 15. $\frac{3}{4}$ 16. 2 17. -1
18. 3 19. $-\frac{10}{3}$ 20. -15 21. -5 22. -6
23. -3 24. 1 25. 5 26. 7 27. -1
28. $7x + 1 = 36$; 5 black keys
29. $3.5x + 812 = 2450$; 468 sandwiches
30. $2x + 24 + 30 = 64$; 5 in.

Reteaching with Practice

1. -2 2. 1 3. 24 4. 2 5. -4 6. 2.5
7. $3 + 1.5n = 12$; 6 8. $6 + 2.5h = 16$; 4

Interdisciplinary Application

1. $c = 112$ chirps per minute 2. $t = 50°F$
3. $t = 67°F$ 4. a. $c = -8$ chirps per minute
b. This answer does not make sense. A cricket cannot chirp a negative number of times.

Challenge: Skills and Applications

1. $1\frac{17}{30}$ 2. $\frac{3}{16}$ 3. $10, -10$ 4. $\frac{22}{3}, -\frac{22}{3}$
5. $29, -29$ 6. $21, -21$ 7. -2 8. $-\frac{7}{2}$
9. $\frac{1}{2}$ 10. $3, -3$ 11. $0.4x + 3 = 15$; Marianna can test 30 plants.
12. $3x + 2(\frac{2}{3}x) + 14\frac{1}{2} = 86$; $16\frac{1}{2}$; three sides are $16\frac{1}{2}$ cm, two sides are 11 cm, and one side is $14\frac{1}{2}$ cm.
13. Sample equation: $25q + 10(q + 2) + 5(2q) + [40 - (q + (q + 2) + 2q)] = 386$, where q = number of quarters; 8 quarters, 10 dimes, 16 nickels, 6 pennies

Quiz 1

1. -8 2. -39 3. $\frac{1}{3}$ 4. 42 5. 4 6. -15
7. 14

Lesson 3.4

Warm-Up Exercises

1. $8x - 10$ 2. $3x + 5$ 3. $-14 + 35x$
4. $8x - 6$ 5. $12x - 4$

Daily Homework Quiz

1. 2 2. 6 3. 15 4. -10 5. 39 hours

Lesson 3.4 *continued*

Lesson Opener

Allow 10 minutes.

1. In the first equation, circle $4x$ and $2x$. In the second equation, circle $10m$ and $2m$. They have variables on each side.   **2. a.** $2x$ was subtracted from each side.   **b.** Like terms are grouped.
c. $2x$; simplify by combining like terms.
d. 5; 5; 5 is subtracted from each side.
e. -8; simplify each side.   **f.** 2; 2; divide each side by 2.   **g.** -4; simplify each side.
 3. a. $10m$ was subtracted from each side.
b. Like terms are grouped.   **c.** $-8m$; simplify by combining like terms.   **d.** -8; -8; divide each side by -8.   **e.** 2; simplify each side.

Practice A

1. a. Subtract $7x$ from each side of the equation.
b. Add 4 to each side of the equation.
c. Divide each side of the equation by 2.
 2. a. Subtract 3 from each side of the equation.
b. Add $4x$ to each side of the equation.
c. Divide each side of the equation by 6.
 3. a. Use the distributive property on the left side of the equation.
b. Add 36 to each side of the equation.
c. Divide each side of the equation by 8.

 4. 9; subtract 9 from each side.   **5.** 6; subtract $4x$ from each side, then divide each side by -1.
6. 8; add $3x$ to each side.   **7.** 12; subtract $5x$ from each side, then divide each side by 2.
8. -5; subtract $7x$ from each side, then divide each side by -1.   **9.** -5; subtract $9x$ from each side, then divide each side by 3.

10. 5   **11.** 10   **12.** 2   **13.** -3 **14.** -7

15. no solution   **16.** identity   **17.** -1 **18.** -2
19. $\frac{22}{3}$ **20.** 7   **21.** -6 **22.** $x + 12 = 2x$;
12 feet   **23.** $6 + 3x = 5x$; 3 grams
24. $40t = 55(t - 3)$; 11 hours

Practice B

1. 3; add x to each side, then divide each side by 3.   **2.** -1; add $4x$ to each side, subtract 1 from each side, then divide each side by 7.
3. $-\frac{3}{5}$; add $2x$ to each side, subtract 8 from each side, then divide each side by 5.   **4.** -11; distribute 4 on the left side, distribute -7 on the right side, add $7x$ to each side, subtract 40 from

each side, and then divide each side by 3.   **5.** 1; distribute $\frac{2}{3}$ on the left side, subtract $6x$ from each side, add 4 to each side, and then divide each side by 2.   **6.** 4; distribute -1 on the left side, distribute 2 on the right side, group $-12x$ and x on the right side, simplify, add $11x$ to each side, add 18 to each side, and then divide each side by 10.

 7. 5   **8.** 12   **9.** 4   **10.** -5 **11.** -5

12. no solution   **13.** identity   **14.** 1   **15.** -11
16. $\frac{23}{3}$ **17.** 7   **18.** -24 **19.** no solution
20. -3 **21.** identity   **22.** $2x + 12 = 4x$; 6 ft
23. $4 + 2x = 8x$; $\frac{2}{3}$ g
24. $35t = 50(t - 3)$; 10 hours; 7 hours
25. Answers vary.
Sample answer: $x + 1 = x + 3$
26. Answers vary.
Sample answer: $4x + 2 = 2(2x + 1)$

Reteaching with Practice

1. $\frac{1}{2}$ **2.** -2 **3.** -1 **4.** no solution
 5. identity   **6.** $-\frac{1}{2}$ **7.** 40   **8.** 55

Learning Activity

1. Answers vary.   **2.** Answers vary.
3. Answers vary.

Real-Life Application

1. $0.48p = 0.42p + 1.8$ **2.** $p = 30$ pounds
3. Earth Saver Recycling Center; You will be paid more at the Earth Saver Recycling Center if you have less than 30 pounds of aluminum cans.
4. Cans-for-Cash Recycling Center; You will be paid more at the Cans-for-Cash Recycling Center if you have more than 30 pounds of aluminum cans.

Challenge: Skills and Applications

1. -24 **2.** 6   **3.** $\frac{49}{3}$ **4.** 12   **5.** identity
6. no solution   **7.** $-\frac{1}{4}$
8. $\frac{11}{12}$ **9.** $-\frac{2}{3}$

Lesson 3.4 *continued*

10.

$$\frac{1}{90} \cdot \boxed{\begin{array}{c}\text{Gross}\\\text{Income}\end{array}} =$$

$$\frac{1}{60} \cdot \left(\boxed{\begin{array}{c}\text{Gross}\\\text{Income}\end{array}} - \boxed{\text{Expenses}} \right)$$

$\frac{1}{90}x = \frac{1}{60}(x - 100,000)$, where $x =$ gross income.
11. \$300,000; the plans are the same when the company has a gross income of \$300,000.
12. \$3333.33 **13.** The first plan is better.

Lesson 3.5

Warm-Up Exercises

1. 5 **2.** −6 **3.** 3 **4.** $\frac{3}{5}$

Daily Homework Quiz

1. 9 **2.** 7 **3.** $\frac{2}{5}$ **4.** 4 **5.** $\frac{1}{5}$ hour or 12 minutes

Lesson Opener

Allow 10 minutes.
1. a. The amount earned each week is \$50.
b. $750 = 50x$ **2. a.** *Sample answer:* The length of the dog run is twice its width.
b. $x + 2x + x + 2x = 30$ or $6x = 30$

Practice A

1. 2 **2.** −1 **3.** −6 **4.** −3 **5.** 17 **6.** $\frac{19}{3}$
7. −18 **8.** $-\frac{2}{5}$ **9.** 0 **10.** $\frac{11}{13}$

11.

$$\boxed{\begin{array}{c}\text{Left and}\\\text{right margins}\end{array}} + 2 \cdot \boxed{\begin{array}{c}\text{Space between}\\\text{pictures}\end{array}}$$

$$+ 3 \cdot \boxed{\begin{array}{c}\text{Width of}\\\text{picture}\end{array}} = \boxed{\begin{array}{c}\text{Page}\\\text{width}\end{array}}$$

12. $3\frac{1}{2} + 2 \cdot \frac{1}{4} + 3x = 8\frac{1}{2}$ **13.** $1\frac{1}{2}; 1\frac{1}{2}$ in. per picture

14.

$$\boxed{\begin{array}{c}\text{Your current}\\\text{savings}\end{array}} + 5 \cdot \boxed{\begin{array}{c}\text{Number}\\\text{of weeks}\end{array}} =$$

$$\boxed{\begin{array}{c}\text{Sister's current}\\\text{savings}\end{array}} - 10 \cdot \boxed{\begin{array}{c}\text{Number}\\\text{of weeks}\end{array}}$$

15. $60 + 5x = 135 - 10x$ **16.** 5; 5 weeks

17.

Week	0	1	2	3	4	5
Your money	\$60	\$65	\$70	\$75	\$80	\$85
Sister's money	\$135	\$125	\$115	\$105	\$95	\$85

Practice B

1. −1 **2.** −1 **3.** $-\frac{5}{3}$ **4.** −2 **5.** $-\frac{6}{7}$
6. $\frac{19}{3}$ **7.** −3 **8.** 1 **9.** $-\frac{1}{2}$ **10.** $\frac{11}{12}$

11.

$$\boxed{\begin{array}{c}\text{Width}\\\text{of tape}\end{array}} \cdot \boxed{\begin{array}{c}\text{Number}\\\text{of tapes}\end{array}} = \boxed{\begin{array}{c}\text{Number}\\\text{of rows}\end{array}} \cdot \boxed{\begin{array}{c}\text{Width}\\\text{of box}\end{array}}$$

12. $\frac{5}{8}t = 2 \cdot 10$ **13.** 32; 32 tapes

14.

$$\boxed{\begin{array}{c}\text{Detroit's current}\\\text{temperature}\end{array}} + 2 \cdot \boxed{\begin{array}{c}\text{Number}\\\text{of hours}\end{array}} =$$

$$\boxed{\begin{array}{c}\text{Atlanta's current}\\\text{temperature}\end{array}} - 3 \cdot \boxed{\begin{array}{c}\text{Number}\\\text{of hours}\end{array}}$$

15. $69 + 2x = 84 - 3x$ **16.** 3; 3 hours

17.

Hour	0	1	2	3
Detroit temperature (°F)	69	71	73	75
Atlanta temperature (°F)	84	81	78	75

18. Yes; the graph shows that both cities have the same temperature at hour 3.

Reteaching with Practice

1. $-\frac{13}{11}$ **2.** $\frac{9}{4}$ **3.** $2\frac{2}{3}$ inches

Real-Life Application

1. $9x$ **2.** $x - 2; 12(x - 2)$
3. $9x = 12(x - 2)$; 8 days
4. $9x + 12(x - 2) = 2118$; 102 days

Challenge: Skills and Applications

1. $0.55x, (0.75)3 + 0.4(x - 3)$ **2.** $0.55x = (0.75)3 + 0.4(x - 3)$; 7 **3.** 7 lb of rice costs the same at both stores. **4.** 3.85; 7 lb of rice costs \$3.85 at both stores. **5.** Rice is cheaper at the first store for less than 7 pounds and at the second store for more than 7 pounds.

6–9. Check students' work. The types of problems for which diagrams can be helpful include layout problems and distance problems, such as the ones in Lesson 3.5.

Lesson 3.6

Warm-Up Exercises

1. 1040 **2.** 14.26 **3.** 22.0 **4.** −15
5. −3.76

Lesson 3.6 *continued*

Daily Homework Quiz

1. -7 **2.** 10 **3.** 5 **4.** $\frac{10}{11}$ **5.** 18

Lesson Opener

Allow 10 minutes.

1. a. Subtract 12 from each side. **b.** Divide each side by -18. No, just enter $\div -18$.
c. No, the solution is a repeating decimal, $5.\overline{7}$.

2. a. Subtract $0.9x$ from each side.
b. Subtract 4.5 from each side. **c.** Divide each side by 5.7, which is stored in memory.
d. Yes, the solution is an integer, -1.
3. Subtract 23 from each side; divide each side by 12; 6.25 **4.** Add 1.5 to each side; divide each side by 2.4; $-3.458\overline{3}$ **5.** Subtract $2x$ from each side; add 126 to each side; divide each side by 8; 21.375 **6.** Add $1.9x$ to each side; add 0.5 to each side; divide each side by 2.4; $-0.958\overline{3}$

Practice A

1. 42.8; 42.85 **2.** -3.1; -3.06 **3.** 63.6; 63.64 **4.** 2.2; 2.21 **5.** 110.3; 110.26
6. 4.6; 4.59 **7.** 2.67 **8.** 1.29 **9.** 3.75
10. 0.17 **11.** -2.63 **12.** -12.33 **13.** -11.8
14. 5.62 **15.** 0.34 **16.** 1.96 **17.** 1.69
18. -0.16 **19.** -0.47 **20.** 0.90 **21.** 6.56
22. -22.29 **23.** -0.21 **24.** -7.83
25. about 2.58 cm **26.** $74.72 **27.** 19.8 sec

Practice B

1. -42.8; -42.85 **2.** -18.3; -18.30
3. 123.6; 123.64 **4.** -45.9; -45.85
5. 108.5; 108.51 **6.** 4.6; 4.59
7. 4.33 **8.** 3.57 **9.** -2.75 **10.** 5.17
11. -6.38 **12.** 6.86 **13.** 3.11 **14.** -4.19
15. 8.33 **16.** -0.88 **17.** -4.33 **18.** 0.53
19. -1.09 **20.** 0.37 **21.** -0.11 **22.** -6.51
23. -1.82 **24.** 0.82 **25.** 8.22 **26.** -0.90
27. $-18 + 41x = 57$; 1.83
28. $53 + 92x = 74x - 88$; -7.83
29. $264x - 32 = 59x - 321$; -1.41
30. $150 **31.** 10.8 sec **32.** $30.61

Reteaching with Practice

1. 0.57 **2.** -0.06 **3.** 0.81 **4.** 1.17 **5.** 112.5
6. 0.11 **7.** $13.91 **8.** $11.67

Interdisciplinary Application

1. $48.74 - 0.335x$ **2.** $54.5 - 0.37x$
3. $48.74 - 0.335x = 54.5 - 0.37x$; about 164.6 years **4.** -6.4 sec; The men's and women's times may not ever be the same.

Challenge: Skills and Applications

1. Germany, 609.4; India, 791.8; Japan, 861.0; Turkey 211.0; U.S., 72.9 **2.** about 11.8 times
3. 0.056 **4.** D
5. $93.6x$, $85.4(x + 0.75)$; $93.6x = 85.4(x + 0.75)$
6. about 7.81; The shuttle makes about 7.81 orbits and the satellite makes about $7.81 + 0.75 = 8.56$ orbits before they first meet.
7. *Sample explanation:* Find the number of minutes of travel by multiplying the number of orbits the shuttle makes (n) by the minutes per orbit it takes (93.6). Then find how many orbits the satellite makes in that time by dividing the number of minutes by the satellite's minutes per orbit (85.4); Solution checks since $(7.81)(93.6) \div 85.4 \approx 8.56$.
8. 12 h, 11 min

Quiz 2

1. -2 **2.** 11 **3.** $\frac{1}{2}$
4. a. $120 + 4x = 8(10 + x)$
b. Plan A: $10 per month; Plan B: $20 per month
5. 0.06 **6.** 3.52

Lesson 3.7

Warm-Up Exercises

1. 7 **2.** -24 **3.** 58 **4.** -8 **5.** 48

Daily Homework Quiz

1. 1.79 **2.** -1.40 **3.** 1.77 **4.** 0.91 **5.** $23.68

Lesson 3.7 *continued*

Lesson Opener

Allow 15 minutes.

1. *Sample answers:* Geometry: $A = \ell w$; Science: $F = \frac{9}{5}C + 32$; Consumer mathematics: Sales tax = Tax rate times Price; Sports: Batting average = Number of hits divided by Number of times at bat. **2.** *Sample answer:* $A = \ell w$, A represents area of a rectangle, ℓ represents length, and w represents width. **3.** *Sample answer:* Area formula is used to find how much material to use to cover a floor; temperature conversion formula is used to compare Celsius and Fahrenheit units of measurement; sales tax formula is used to find the amount of tax charged on a purchase price; baseball statisticians keep records on all baseball players and compare their batting averages.

Practice A

1. $w = \dfrac{A}{\ell}$ **2.** $r = \dfrac{C}{2\pi}$ **3.** $s = \dfrac{P}{4}$ **4.** $r = \dfrac{d}{t}$

5. $R = \dfrac{E}{I}$ **6.** $h = \dfrac{3V}{\pi r^2}$ **7.** $y = 3x + 8$

8. $y = 2x + 15$ **9.** $y = -7x + 4$

10. $y = -2x - 4$ **11.** $y = -x + 12$

12. $y = 3x + 2$ **13.** $y = -8x + 2$

14. $y = 6x + 14$ **15.** $y = -\frac{2}{3}x + 5$

16. $y = \frac{2}{5}x - 1$ **17.** $y = 2x - \frac{5}{2}$

18. $y = 2x - 5$ **19.** $x = y + 4; 2; 3; 4; 5$

20. $x = 2y - 3; -7; -5; -3; -1$

21. $x = -2y + 3; 7; 5; 3; 1$ **22.** $t = \dfrac{d}{r}$

23. 10 hours; 5 hours; $3\frac{1}{3}$ hours

24. $r = \dfrac{I}{Pt}$ **25.** 5.5%

Practice B

1. $t = \dfrac{I}{Pr}$ **2.** $d_2 = \dfrac{2A}{d_1}$ **3.** $b_1 = \dfrac{2A - b_2 h}{h}$

4. $C = \dfrac{5}{9}(F - 32)$ **5.** $\ell = \dfrac{2(S - B)}{P}$

6. $h = \dfrac{S - 2\pi r^2}{2\pi r}$ **7.** $y = -9x + 4$

8. $y = \frac{2}{5}x + 3$ **9.** $y = 5x - 4$

10. $y = -3x - 2$ **11.** $y = -2x + 8$

12. $y = -3.5x + 7$ **13.** $y = -48x + 12$

14. $y = -6x + 14$ **15.** $y = -\frac{2}{3}x + 5$

16. $y = \frac{2}{5}x - 1$ **17.** $y = -2x + \frac{9}{2}$

18. $y = -4x - 5$ **19.** $y = \dfrac{14 - 7x}{3}$

20. $y = -2x + 8$ **21.** $y = \dfrac{-2x + 5}{-3}$

22. $x = 2y - 3; -7; -5; -3; -1$

23. $x = \frac{1}{5}y + 2; \frac{8}{5}; \frac{9}{5}; 2; \frac{11}{5}$

24. $x = \frac{1}{2}y + 1; 0; \frac{1}{2}; 1; \frac{3}{2}$

25. $x = -3 - y; -1; -2; -3; -4$

26. $x = -\dfrac{1}{2}y - 1; 0; -\dfrac{1}{2}; -1; -\dfrac{3}{2}$

27. $x = \dfrac{4y - 16}{3}; -8; -\dfrac{20}{3}; -\dfrac{16}{3}; -4$

28. $t = \dfrac{d}{r}$

29. 12 hours 30 minutes; 6 hours 15 minutes; 4 hours 10 minutes

30. $P = \dfrac{I}{rt}$ **31.** $1375

Reteaching with Practice

1. $h = \dfrac{2A}{b}$ **2.** $r = \dfrac{C}{2\pi}$ **3.** $P = \dfrac{I}{rt}$ **4.** $r = \dfrac{I}{Pt}$

5. $r = \dfrac{d}{t}$ **6.** 58 mi/h

Interdisciplinary Application

1. 3 **2.** 2.5 **3. a.** $h_i = Mh_o$ **b.** 21 in.

4. a. $d_o = \dfrac{d_i}{M}$ **b.** 12 cm

5. The image is smaller than the actual object.

Challenge: Skills and Applications

1–5, 7. Accept equivalent forms.

1. $x = \frac{7}{6}(p - 2q)$ **2.** $x = \frac{10}{3}ab - 10$

3. $x = h - \dfrac{1}{10}k$ **4.** $x = \dfrac{8r + 4t - 8}{5}$

5. $d = 0.86p - 5; p = \dfrac{d + 5}{0.86}$ **6.** about $15.26

7. $d = 0.86(p - 5); p = \dfrac{d}{0.86} + 5$

Lesson 3.7 *continued*

8. about $14.44

9. $y = 1200 + 0.05(x - 30,000)$, $y = 2700 + 0.07(x - 60,000)$

10. $x = \dfrac{y}{0.05} + 6000$, $x = \dfrac{y}{0.07} + 21,429$

11. $65,000

Lesson 3.8

Warm-Up Exercises

1. 0.8; 80%   **2.** 2; 200%   **3.** 1.125; 112.5%
4. $\dfrac{7}{8}$ **5.** $\dfrac{5}{7}$

Daily Homework Quiz

1. $a = \dfrac{F}{m}$ **2.** $y = \dfrac{x + 5}{3}$ **3.** 3   **4.** 5

5. $t = \dfrac{A - 10,500}{18}$; 21 minutes

Lesson Opener

Allow 15 minutes.

1. $\dfrac{11}{9}$; 105th Congress   **2.** $\dfrac{12}{13}$; Find the reciprocal of $\dfrac{12}{13}$.   **3.** 56%; 44%   **4. a.** 14 students per teacher   **b.** 14 students per teacher   **c.** 14 students per teacher   **d.** 13 students per teacher   **e.** 13 students per teacher

Practice A

1. $0.40 per can   **2.** $22.50 per ticket
3. about $0.24 per apple   **4.** $3\frac{1}{2}$ cups per loaf
5. 8 oz per serving   **6.** 12 bottles per carton
7. 390 miles per hour   **8.** 480 miles per day
9. 4.8 miles per hour   **10.** 46 pages per hour
11. 0.075 km per min   **12.** 1.5 acres per hour
13. 1500 tickets per hour   **14.** $3\frac{1}{6}$ inches per hour
15. 41.2 miles per hour   **16.** $\dfrac{160}{1}$
17. about 25.8 miles per gallon   **18.** 387 miles

Practice B

1. $0.80 per can   **2.** $21.75 per ticket
3. about $0.25 per apple   **4.** $7.25 per hour
5. 6 oz per serving   **6.** about $1.99 per quart
7. 375 miles per hour   **8.** $15\frac{1}{3}$ miles per day

9. 0.08 mile per minute   **10.** $\dfrac{3}{4}$ acre per hour
11. 43 words per min   **12.** 104 km per hour
13. 1595 tickets per hour   **14.** 3.07 inches per hour   **15.** 23 miles per hour   **16.** $\dfrac{160}{1}$
17. about 25.7 miles per gallon   **18.** 385.5 miles

Reteaching with Practice

1. $\dfrac{24}{1}$ **2.** 40 mi/h   **3.** $11/h   **4.** about 190
5. about 310   **6.** about 284 marks
7. about 2665 schillings

Learning Activity

1. Answers vary.   **2.** *Sample answer:* The sides of a Golden Rectangle have "normal" proportions, and forms of the Golden Rectangle are seen in all forms of nature, art, and architecture.   **3.** *Sample answer:* The height of a spruce tree and its width (measuring the branches at the bottom) are in the Golden proportion. Features of this proportion are also found in the human face.

Real-Life Application

1. $88x = 1483$; about 16.85 feet
2. $\dfrac{13}{10}$ **3.** $\dfrac{19}{44}$ **4.** $\dfrac{4}{15}$

Challenge: Skills and Applications

1. about 27.3 mi/h   **2.** about 1.0 kg/L
3. 0.3 mm/min   **4.** 0.1 mi/min   **5.** 3.4 lb/ft
6. 58,200 ft/min   **7.** 5 cm/h   **8.** about $2.12/gal

Lesson 3.9

Warm-Up Exercises

1. 24%   **2.** 316%   **3.** 80%   **4.** 33.3%

Daily Homework Quiz

1. $\dfrac{3}{4}$ **2.** $\dfrac{3}{2}$ **3.** $0.25 per can   **4.** 72 inches
5. 1.2 miles per minute

Lesson 3.9 *continued*

Lesson Opener

Allow 15 minutes.

1, 2.

	Percent	Percent as a decimal	Number n
Sample	15%	0.15	20
Row 1	25%	0.25	12
Row 2	50%	0.50	24
Row 3	75%	0.75	16
Row 4	80%	0.80	20

	Percent · n
Sample	$(0.15)(20) = 3$
Row 1	$(0.25)(12) = 3$
Row 2	$(0.50)(24) = 12$
Row 3	$(0.75)(16) = 12$
Row 4	$(0.80)(20) = 16$

3. $0.25b = 3$; divide each side of the equation by 0.25. **4.** $p \cdot 24 = 12$, or $24p = 12$; divide each side of the equation by 24.

5. $0.75(16) = a$; simplify the left side of the equation.

Practice A

1. 0.28 **2.** 0.5 **3.** 0.8 **4.** 0.15 **5.** 0.04

6. 0.085 **7.** 1.22 **8.** 0.005 **9.** 9 **10.** 12%

11. 600 **12.** 160 **13.** 120 **14.** 64%

15. 88% **16.** 80 **17.** 180 **18.** 72 **19.** 60%

20. 120 **21.** $33\frac{1}{3}\%$ **22.** 50% **23.** $33\frac{1}{3}\%$

24. 75% **25.** 16 **26.** 24 **27.** 98 **28.** 62

Practice B

1. 219 **2.** 138.6 **3.** 44% **4.** 70 **5.** 240

6. 60% **7.** 80% **8.** 220 **9.** 155 **10.** 650

11. 160% **12.** 2.4 **13.** about 37%; about 63%

14. about 44%; about 56%

15. about 1.90933×10^8 mi^2

16. about 2.487×10^8 **17.** 75% **18.** 50%

19. 200% **20.** 600%

Reteaching with Practice

1. $7.20 **2.** 9 mi **3.** 150 g **4.** 50 yd

5. 25% **6.** 400% **7.** 170

Interdisciplinary Application

1. $0.20 **2.** $7; $27 **3.** $10; 40% **4.** 28.6%

5. $8; 87.5%

Challenge: Skills and Applications

1. $66\frac{2}{3}\%$ **2.** $\frac{1}{5}$ **3.** $\frac{75}{38}$ **4.** 56% **5.** 157.5 in.^{2}

6. $\frac{9}{16}$ **7.** The area is changed by a factor of p^2.

8. 0.6; 1.2; 0.72 **9.** $\left(1 - \frac{x}{100}\right)\left(1 + \frac{y}{100}\right)$ or $\left(\frac{100 - x}{100}\right)\left(\frac{100 + y}{100}\right)$ **10.** 20%

Review and Assessment

Review Games and Activity

1. $x = -9$ **2.** $x = -30$ **3.** $x = 11$

4. $x = 8$ **5.** $x = -6$ **6.** $x = -95$

7. $x = 2$ **8.** $x = -3$ **9.** $x = 19$

10. $x = 56$ **11.** $x = 24$ **12.** $x = 32$

13. $x = 4$ **14.** $x = -2$ **15.** $x = 7$

16. $x = 10$ **17.** $x = -56$ COOL AID

Chapter Test A

1. 25 **2.** -25 **3.** $26 **4.** yes **5.** no

6. -2 **7.** 28 **8.** $28.75 each week

9. 128 **10.** 3 **11.** 138, 139, 140; $x + (x + 1) + (x + 2) = 417$ **12.** 2

13. 2 **14.** $10x = 300 + 4x$; 50 hours **15.** 136 pixels **16.** $-32.1, -32.06$ **17.** $-37.7, -37.69$

18. 3.13 **19.** 5.86 **20.** -4.33 **21.** 11.44

22. 1.42 **23.** 0.22 **24.** $b_1 = \frac{2A}{h} - b_2$

25. $t = \frac{I}{Pr}$ **26.** $y = 13 - 7x$ **27.** $y = 3 - \frac{3}{5}x$

28. 20, 13, -1 **29.** $\frac{18}{5}, 3, \frac{9}{5}$

30. 2 tablespoons per serving

31. $0.92 per slice **32.** 5 tablespoons **33.** 45%

Chapter Test B

1. 7 **2.** 16 **3.** $22 **4.** no **5.** yes **6.** 32

7. 35 **8.** 18 **9.** 5 **10.** 8 **11.** 4 **12.** 6

Review and Assessment *continued*

13. 23, 50, 64 **14.** identity **15.** 5

16. $8x = 150 + 2x$; 25 days **17.** $-5.7, -5.65$

18. 41.7, 41.68 **19.** 1.07 **20.** -13.54

21. 0.25 **22.** 4.43 **23.** $h = \dfrac{3V}{\pi r^2}$

24. $C = \frac{5}{9}(F - 32)$ **25.** $y = 3x - 6$

26. $y = -2x - 5$ **27.** $-9, -6, 0$

28. $-3, -5, -9$

29. 32 ounces for $2.40

30. 236.5 kilometers **31.** $333.33

SAT/ACT Chapter Test

 1. A **2.** D **3.** C **4.** C **5.** D **6.** D **7.** A

 8. B **9.** B **10.** C **11.** D **12.** C

Alternative Assessment

 1. Complete answers should include these points:

$7 + \frac{1}{2}(6x + 4) = -2x + 4$	Original equation
$7 + 3x + 2 = -2x + 4$	This step is correct.
$10x + 2 = -2x + 4$	Error: added 7 and 3x to get 10x; incorrect because 7 and 3x are not like terms. The correct method is to add like terms 7 and 2, to get $3x + 9$.
$8x + 2 = 4$	Error: incorrect to add 10x and $-2x$; because they are on different sides of equation. The correct method is to add 2x to *both* sides of the equation.
$8x = 2$	This step is correct.
$x = 4$	Error: did not divide both sides by 8 to isolate the variable. The correct method is to divide *both* sides by 8 to isolate the variable.

 2. a. $3350; $2700; Salary Plan 1

 b. total pay = base pay +
 percent commission · sales in a month

 c. total pay = percent commission ·
 sales in a month

 d. $2000 + 0.03x = 0.06x$; Answers should include these points: If Ron's sales per month are $66,666 or less, he should choose Plan 1. If his sales per month are greater than $66,667, he should choose Plan 2. **e.** Ron should chose Plan 2 because monthly sales must exceed $66,667 in order to provide a salary of $8000 a month. Since $8000 = 0.06x$ (where x = sales per month), $x = 133,333$. Ron must make $133,333 in sales each month.

 3. 15 houses under Plan 1; 14 houses under Plan 2

Project: Ice Rescue

 1. Make sure students determine the area of the bottom of the can correctly and divide the total weight by this area to find the stress. *Sample answer:* about 5 pounds per square inch

 2. Make sure area estimates are reasonable. Equations should be of the form

$$\frac{x + 100}{2 \times \text{area of foot}} = \text{the stress the ice can bear.}$$

Sample answer: 32 in.^{2}; $\dfrac{x + 100}{64} = 5$; 220 lb

 3. The equation should be of the form $\dfrac{x + 102}{4} = $ the stress the ice can bear.

Sample answer: $\dfrac{x + 102}{4} = 5$; -82 lb

 4. The equation should be of the form $\dfrac{x + 110}{2592} = $ the stress the ice can bear.

Sample answer: $\dfrac{x + 110}{2592} = 5$; 12,850 lb

Cumulative Review

 1. 49 **2.** 219.4 **3.** 18 **4.** 25 **5.** yes
 6. yes **7.** no **8.** yes **9.** yes **10.** no

Review and Assessment *continued*

11. $4 > 5x$ **12.** $8x - 12 < 50$ **13.** $\frac{1}{2}x \geq 100$

14. $2\frac{5}{9} - x = 31$ **15.** false **16.** true

17. false **18.** true **19.** $-6, -\frac{10}{3}, 0, \frac{12}{8}, 2, 3$

20. $-8.9, 8.023, 8.06, 8.5, 8.599, 8.6$

21. $-3\frac{6}{7}, -3.6, -3.023, -3, 3.25, 3\frac{2}{5}$

22. $-9.01, -8.9, -\frac{34}{9}, -1, -\frac{4}{17}, 9$ **23.** 5.03

24. $9\frac{5}{8}$ **25.** 9.569 **26.** $1\frac{5}{24}$ **27.** -9.01

28. -6.36 **29.** 11 **30.** $\frac{1}{2}$ **31.** $3\frac{14}{15}$

32. -0.6 **33.** 0 **34.** -7.2 **35.** -51

36. 0 **37.** $\frac{1}{6}$ **38.** 108 **39.** $-\frac{119}{48}$ **40.** 6

41. no **42.** no **43.** no **44.** yes

45. no solution **46.** no solution **47.** 245

48. no solution **49.** $-\frac{2}{5}$ **50.** $\frac{29}{18}$

51. \$28.75 **52.** \$21.20

53. 1.5 **54.** -0.1 **55.** 19.6 **56.** 2.6

57. $y = \frac{5}{2}x - 4$ **58.** $\frac{1}{2}x - \frac{9}{2} = y$

59. $\dfrac{2x + 7}{3}$ **60.** $y = \frac{15}{2}x - 1680$

61. 32 weeks **62.** 243.8 cm **63.** 2.4 km

64. \$28.22 U.S.

Cumulative Test

1. 16 **2.** 49 **3.** 512 **4.** 15 **5.** -288 **6.** 2

7. 64 **8.** 2 **9.** 2 **10.** 24 **11.** -27

12. $-\dfrac{5}{3}$ **13.** -1 **14.** $x + 7 = 23$

15. $x - 12 = 16$ **16.** $35 > 5x$ **17.** $\dfrac{x}{6} < 27$

18.

Input	Output
0	-5
1	-2
2	1
3	4

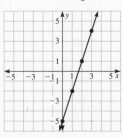

19.

Input	Output
0	8
1	4
2	0
3	-4

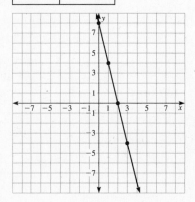

20. $-3.9, -1.72, -1.4, 0.6, 2$

21. $-2, -1.7, -\dfrac{4}{3}, -0.5, \dfrac{3}{2}$

22. $-5.3, -5, 1, |2.7|, |-6|$ **23.** $-8m$

24. $c + 5$ **25.** $3p^2 + 4p - 6$ **26.** $14r - 11$

27. $4x - 46$ **28.** $3x - 21$ **29.** $-30 - 6a$

30. $-18a - 36$ **31.** -1 **32.** 7

33. 12 **34.** 2 **35.** -4 **36.** 7 **37.** -4

38. $y = z - x$ **39.** $x = \dfrac{(c + b)}{a}$

40. $m = \dfrac{(n - 4)}{4}$ **41.** \$112.50 **42.** 28.8 feet

43. $33\frac{1}{3}$ grams **44.** \$666.67 **45.** 9% **46.** 14%

47. $5(2 + 0.30) = \$11.50$

48. $25 - 17 = x$; 8 people were left

49. $150 + 30x = 360$; 7 weeks